WINKS
User's Guide

Statistical Software for Windows
Fifth Edition

For updates, errata and other information concerning this manual or program go to
http://www.texasoft.com/updates

WINKS

The WINKS software and manual ("documentation") are copyrighted by TexaSoft and are protected by both United States copyright and International treaty provisions. Use of the software indicates your acceptance of the following License, Limited Warranty and Governing Law statements.

LICENSE STATEMENT: WINKS ("this software") is licensed for use on one computer only at any time. This software is not copy protected. TexaSoft authorizes you to make back-up copies of the software for your protection.
This software is just like a book, which cannot be read by two people at two different locations at the same time (unless, the copyright has been violated.) If you wish to run several versions of this program on different computers at one time, you need to pay a license fee for each computer. If you have one computer at work, and another at home, and they will not be used at the same time, you can load the program on both computers without purchasing an additional license.

LIMITED WARRANTY: TexaSoft does not warrant that the functions contained in this program will meet your requirements or that the program operation will be uninterrupted or error free. This program is sold to you on an "as is" basis. The only exception to this disclaimer is a 30-day replacement or refund warranty. In no event shall TexaSoft be liable or responsible for any problems that arise due to defects with this software.

GOVERNING LAW: These statements shall be construed, interpreted, and governed by the laws of the state of Texas. You agree that this is the complete and exclusive statement of this agreement which supersedes any prior agreement or other communication between us on this subject. Use, duplication, or disclosure by the U.S. Government of the computer software and documentation in this package shall be subject to the restricted rights under DFARS 52.227-7013 applicable to commercial computer software.

Contents

Part 1 – Overview and Tutorials 9

Getting Started 9
An Overview of WINKS 9
Installation 10
Tutorial 1 – Analyzing Summary Data....................................11
Comments on Tutorial 1 – Analyzing Summarized Data..........12
Tutorial 2 – Entering Data to Analyze – Comparing Means 13
Tutorial 3 – Using Data from an Existing File 19
Tutorial 4 – Using WINKS GRAPHS 22
Tutorial 5 – Importing Data from Excel 26
Tutorial 6 – Pasting output into Word 28

Part 2 – Data Topics.. 29

WINKS Data Topics..29
Creating a New Database..30
    Creating a Database from a Pre-defined Structure 30
    Creating a Customized Database 31
Customized Database Example 32
WINKS Database Limitations 35
File Menu...35
Edit Menu 36
Spreadsheet File and Edit Menus 36
Data Editing Techniques...40
    Missing Values Codes 40
    Sorting..41
    Selecting Random Samples 41
    Formula Entry and Text Replacement...............................42
    Data Editor Recode...43
Entering and Importing Data 44

Entering Data From an ASCII..44
Appending Data from Another dBase.............................44
Moving Data from Another Program............................44
Excel...45
Importing Nondelimited ASCII 46
Importing Comma Delimited 46
Printing a Data Report...47

Part 3 – Statistical Concepts Review 49

Using Statistics to Analyze Information 49
Summarizing Information with Statistics 51
Describing Quantitative Data...52
Using Histograms to Examine Data...57
Selecting a Measure of Dispersion.......................................58
Describing Qualitative Data...59
Describing a Linear Relationship Between Quantitative
Variables 61
Describing a Relationship Between Qualitative Variables ... 64
Using Statistics to Make Comparisons 66
Performing a Statistical Test 67
Stating the Hypotheses...68
Performing the Analysis...69
Interpreting the Test and Making a Decision......................71
Interpreting Multiple Comparisons.....................................73
Choosing the Right Procedure to Use 75
Descriptive Statistics...76
Comparison Tests (t-test/ANOVA) 77
Relational Analyses (Correlation and Regression) 78

Part 4 – BASIC Statistical Procedures......79

Detailed Statistics and Histogram (Analyze/Descriptives/Detail-
One Variable)..80
Normal Probability Plot 82
Detailed Statistics With Cp and Cpk
(Analyze/Descriptives/Detail Cp & Cpk) 82

Summary Statistics on a Number of Variables 83
Percentile Calculation (Analyze/Descriptives/Percentile
Calculation) 84
Detailed Statistics from Data Entered by Counts
(Analyze/Descriptives/Detailed Statistics from Counts) 84
Stem and Leaf Display (Analyze/Descriptives/Stem and Leaf
Display) 85
Approximate p-value Determination (Analyze/Descriptives/p-
Value Determination) 86
  Probability Calculator 87
Graphs and Charts 88
Histogram/Stats (Analyze/Graphs/Charts – Histogram/Stats) .. 88
XY Plot (Scatterplot) 88
  Line Chart (Analyze/Graphs/Charts – Line Chart) 89
  Correlation Matrix Graphs (Analyze/Graphs/Charts –
  Correlation Matrix Graphs) 89
  By Group Plot (Analyze/Graphs/Charts – By Group Plots) . 90
  By Group Plots allow you to compare distributions of data by
  groups. The following example compares AGE by GROUP in
  the EXAMPLE database. 90
  First Impression Plots (Analyze/Graphs/Charts – First
  Impression Plots) 91
  First Impression X-Y Scatterplot 91
  First Impression Bar Chart 91
  First Impression Pie Chart 92
  First Impression Line/Area Chart 93
t-Tests and ANOVA 94
Independent Group t-test (Analyze/t-Tests and ANOVA –
Independent Group t-test/ANOVA) 94
  One-Way ANOVA Example (Analyze/t-Tests and ANOVA
  – Independent Group t-test/ANOVA) 97
Independent group test from summary data (Analyze/t-Tests
and ANOVA – Ind. Group from Summary Data) 100
  Paired and Repeated Measures Analyses 101
  Paired t-test Example (Repeated Measures) (Analyze/t-Tests
  and ANOVA – Paired Rep. Measures (t-test/ANOVA)) 102

One-way repeated measures ANOVA Example (Analyze/t-Tests and ANOVA – Paired Rep. Measures (t-test/ANOVA)) 104
Single Sample t-test Analysis (Analyze/t-Tests and ANOVA – Single Sample t-test) 106
Single Sample t-test/Summary Data (Analyze/t-Tests and ANOVA – Single Sample t-test Summary Data) 107
Dunnett's Test Single Sample t-test Analysis 107
Dunnett's Test Single Sample t-test Analysis from Summary Data 107
Non-Parametric Procedures 108
Mann-Whitney Example (Analyze/Non-Parametric Comparisons/Ind. Gp. Non-Parametric) 108
Kruskal-Wallis Example (Analyze/Non-Parametric Comparisons/Ind. Gp. Non-Parametric) 109
Wilcoxon Signed Rank Test – Paired Data (Analyze/Non-Parametric Comparisons/Paired Rep. Measures Non-Parametric) 112
Friedman's Test - Repeated Measures (Analyze/Non-Parametric Comparisons/Paired Rep. Measures Non-Parametric) 112
Regression and Correlation 116
Simple Linear Regression Example (Analyze/Regression and Correlation/Simple Linear Regression) 117
Prediction Intervals in Simple Regression 120
Regression Through the Origin 120
Multiple Regression Example (Analyze/Regression and Correlation/Multiple Linear Regression) 121
Correlation Analysis (Analyze/Regression and Correlation/Correlation, Pearson & Spearman) 124
Correlation Matrix (Analyze/Regression and Correlation/Correlation Matrix) 126
Graphical correlation matrix (Analyze/Regression and Correlation/Graphical Correlation Matrix) 128
Point Bi-serial Correlation 129
Analysis of Count Data – Frequencies and Crosstabulations . 130

Goodness-of-Fit Analysis (Analyze/Crosstabs, Frequencies, Chi-Square/Goodness-of-fit) 131
Entering Data from the Keyboard (Analyze/Crosstabs, Frequencies, Chi-Square/Chi-Square from Keyboard)........133
Entering Data from a Database (Analyze/Crosstabs, Frequencies, Chi-Square/Crosstabulations/Chi-Square) 134
Notes on 2x2 Table Statistics 137
Example: Crosstabulation– Homogeneity Hypothesis (Analyze/Crosstabs, Frequencies, Chi-Square/Crosstabulations/Chi-Square).....................................139
McNemar's Test (Analyze/Crosstabs, Frequencies, Chi-Square/McNemar's Test)...140
Comparison of Proportions (Analyze/Crosstabs, Frequencies, Chi-Square/Proportions Comparison) 142
Life Table and Survival Analysis 143
Actuarial Life Table Analysis (Analyze/Life Tables & Survival Analysis/Actuarial Life Table)...........................144
Kaplan-Meier Analysis (Analyze/Life Tables & Survival Analysis/Kaplan-Meier Analysis) 146
Analyze From Summary Data 147
Simulations and Demonstrations 147

Part 5 – Professional Edition Procedures 150

Advanced Tabulation Procedures 150
Advanced Tabulation..150
What are Tabulation Variables? 151
Tabulation Example...156
Defining Format Definitions for Tables 158
Mantel Haenszel Comparison (Analyze./ Advanced Tabulation, and the Mantel-Haenszel Comparisons)..................................159
Inter-Rater Reliability/KAPPA (Analyze/Advanced Tabulation/InterRater Reliability/KAPPA) 162
Advanced ANOVA..163
Two-Way ANOVA (Analyze/Advanced ANOVA/Two-Way ANOVA)..164
Interaction Plots 168

Two-Way Unbalanced Design 168
Two Way Repeated Measure ANOVA (Analyze/Advanced
ANOVA/Tw0-Way Repeated Measures) 169
Advanced Regression Procedures 174
Polynomial Regression (Analyze/Advanced
Regression/Polynomial Regression) 174
All Subsets Regressions (Analyze/Advanced Regression/All
Possible Subsets) 182
Simple Logistic Regression (Analyze/Advanced
Regression/Logistic Regression) 185
    Using Raw Data In Logistic Regression 188
Bland-Altman Plots (Analyze/Advanced Regression and
Comparison/Bland-Altman) 189
Time Series Analysis (Analyze/Time Series Analysis) 192
    Description of the Time Series Process 192
    How To Analyze Time Series Data 193
    Notes on Time Series Analysis 200
    Output Options 200
    Differencing the series 200
    Output Forecast Information to a File 201
Quality Control Charts 202
    WINKS' QC Features 202
X-Bar and R-Charts (Analyze/Quality Control/X-Bar and R-
Charts) 205
X-Bar and S-Charts (Analyze/Quality Control/X-Bar and R-
Charts) 209
EWMA Calculations (Analyze/Quality Control/EWMA) 209
P-Charts (Analyze/Quality Control/P-Charts Equal or Unequal
Subgroups) 210
Control Charts for Individual Measurements 213
Pareto Charts Analysis (Analyze/Pareto Charts) 213

References **217**

INDEX **219**

Part 1 – Overview and Tutorials

Getting Started

- An Overview of WINKS
- Installation
- Tutorial 1 – Analyzing Summary Data
- Tutorial 2 – Entering Data to Analyze
- Tutorial 3 – Using Data from an Existing File
- Tutorial 4 – Using WINKS Graphs
- Tutorial 5 – Importing data from Excel

✦WINKS Tip – Do the tutorials first – it makes running WINKS quicker and easier.

An Overview of WINKS

Managers and researchers must make decisions. Data analysis is a tool that allows you to make informed decisions. WINKS (**WIN**dows **KwikStat**) is a statistical data analysis program designed by professional statistical consultants and researchers to allow you to quickly and easily calculate the most commonly used statistical data analysis procedures and graphs. A few of the main benefits of WINKS include:

1. WINKS provides a help procedure that guides you to the most appropriate analysis for your data. *(See page 67.)*

2. WINKS provides an easy to use data entry system as well as the ability to automatically read and write industry standard DBF (dBASE III and IV) files. In addition, WINKS can also import or paste data from Excel, 1-2-3, ASCII and other types of files.

3. WINKS anticipates your analysis needs and provides sufficient follow up for a particular analysis for you to make a reasonable conclusion -- unlike other statistical programs that make you run each part of an analysis separately and piece your answers together. For example, in the case of Analysis of Variance procedures, multiple comparisons are automatically performed to show you where specific pair-wise differences lie.

4. WINKS often provides interpretation of your analysis. These may be in the form of warnings when common assumptions are not met and in suggestions about how a particular outcome could be interpreted. For most tests, an explicit hypothesis is stated to help you understand the exact intention of the analysis.

Installation

Place the WINKS distribution CD in your CD Drive. This should automatically start the WINKS Installation procedure. If the installation does not automatically begin: click START, then Run... and enter D:\SETUP.EXE (use your CD drive name here in place of the D: if necessary.)

Follow the installation instructions on the screen. When you begin WINKS for the first time, a dialog box will appear asking for a key code. Enter the code provided with your CD. You will also be asked to choose program options. Select options you want to change, or leave the default options in place and choose Ok. You can change these options at any time by choosing "Change Setup Options" from the Help menu.

Tutorial 1 – Analyzing Summary Data

This tutorial shows you how to perform an Independent Group Student's t-test when you have summarized data. For example, suppose you have the following information about two groups you wish to compare:

> Data for Group 1
>
> > Mean = 23.44, Standard Deviation = 3.41, N = 8
>
> Data for Group 2
>
> > Mean = 31.97, Standard Deviation = 3.22, N = 9

Follow these steps to perform an Independent Group t-test:

Step 1: Begin WINKS. From the Analyze menu, **select "t-tests and ANOVA"** and then select the 3rd option -- **"Ind. Group from Summary Data."**

Step 2: When prompted to enter the number of groups, **enter 2.** This dialog box appears:

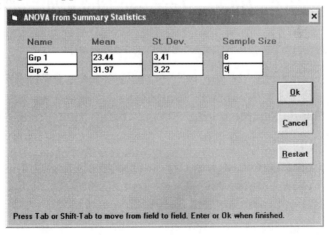

Step 3: Enter the numbers as shown in the dialog box and **Click Ok**. The following output appears. Briefly, this test shows a

significant difference between means of the two groups.

```
--------------------------------------------------------------------------
Independent Group Analysis                                   Summary Data
--------------------------------------------------------------------------

Group Means and Standard Deviations

Grp 1: mean = 23.44            s.d. = 3.41            n =  8

Grp 2: mean = 31.97            s.d. = 3.22            n =  9


Independent group t-test on Summary Data
--------------------------------------------------------------------------
Test equality of variance: F = 1.12 with (7, 8) D.F.  p = 0.866 (two-tail)

Note: Since the p-value for equality of variance is greater than 0.05,
use the Equal variance t-test results.


Independent Group t-test Hypotheses:
------------------------------------
Ho: There is no difference between means.

Ha: The means are different.


  Equal variance: Calculated t= -5.3 with  15  D.F.  p <= 0.001 (two-tail)
Unequal variance: Calculated t= -5.28 with 14.5 D.F. p <= 0.001 (two-tail)
```

Comments on Tutorial 1 – Analyzing Summarized Data

Analyses you can run from summary data include:

- Independent Group Means Comparison – Tests and Analysis of Variance
- Single Sample t-test
- Dunnett's Test
- Contingency Table – Chi-Square Analysis -- also including Yate's Test, Fisher's Exact Test, Relative Risk, Odds Ratio, Sensitivity and Specificity, and more.
- Goodness-of-Fit Analysis
- Test of a Difference between Two Proportions
- P-value determination

Tutorial 2 – Entering Data to Analyze – Comparing Means

When you have the raw data for an analysis, you'll need to enter the data into WINKS in order to perform an analysis. This example shows you how to enter data into WINKS and use that information to perform an analysis and display a graph. Suppose you are testing the effectiveness of four kinds of hog feed. You randomly assign 15 hogs to 4 groups, feed each group one of the feeds for a month, and observe weight gain. You want to know which type of feed produced the most (average) weight gain for the four groups.

GROUP (Feed)	OBS (WEIGHT)
1	60.8
1	67.0
1	54.6
1	61.7
2	78.7
2	77.7
2	76.3
2	79.8
3	92.6
3	84.1
3	90.5
4	86.9
4	82.2
4	83.7
4	90.3

To perform this analysis:

1. Enter the data into a WINKS database
2. Select the type of analysis to perform
3. Observe and interpret the output

Step 1: From the WINKS File menu, **choose "New Database."**
You are asked to enter the name for your new database. **Enter
HOG.** Choose <u>O</u>**k.**

Step 2: - A dialog box appears listing database types. **Click the
down arrow** at the right of the text field to display the options. See
the figure below.

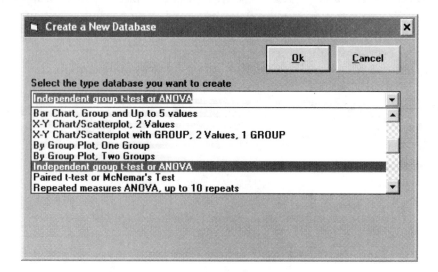

To create a database for this analysis, **choose the option
"Independent Group t-test or ANOVA"** option from the list and
choose <u>O</u>**k.** (This is an analysis used to compare 3 means.)

Step 3: An empty spreadsheet appears containing the two fields
(columns): GROUP and OBS. (See the figure below.) The
GROUP column contains codes for feed type.

```
File  Edit  Options
Data Spreadsheet
        GROUP     OBS
  1        1     60.80
  2        1     67.00
  3        1     54.60
  4        1     61.70
  5        2     78.70
  6        2     77.70
  7        2     76.30
  8        2     79.80
  9        3     92.60
 10        3     84.10
 11        3     90.50
 12        4     86.90
 13        4     82.20
 14        4     83.70
 15        4     90.30
```

For this example we'll use 1, 2, 3, and 4 as the Group names/Feed codes – you could use A, B, C, D or any alpha-numeric name for the group names. The OBS field contains the variable you observed. In this case it is weight gain.

Enter the data into the spreadsheet as shown in the figure above. When you enter a complete row, and press Enter and a new blank row will be added to the spreadsheet.

Note:
You should not press enter on the last row, so you won't end up with a row of blank cells. If you do end up with a blank row, click on it, then choose "Delete Row" from the Edit menu to delete that row before saving the database

Step 4: When you have finished entering the data, **select File and "Exit Edit."** You will be asked if you want to save the data before exiting. **Answer "Yes."**

Step 5: From the main WINKS menu screen **click on the Analyze menu** and **select the "t-test and ANOVA"** option. A sub menu will appear. From the sub-menu, **choose the "Independent Group (t-test/ANOVA)** option.

Step 6: Choose what fields to use. A dialog box appears allowing you to select which fields to use for this analysis. **Select the GROUP** field, and click on the **Group** button. Then **select the OBS field** and click on the **Add** button. Your field choices will look like the dialog box in the figure below. **Click Ok** to continue.

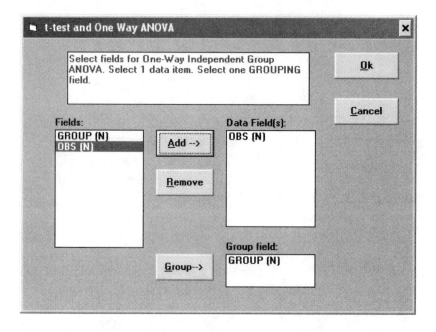

Step 7: The following output information is displayed. Use the scroll bars to view output and/or print the output by clicking on the "Print" button. For interpretation of this output, refer to the WINKS manual.

Briefly, this analysis compares the 4 means and shows that feed types 3 and 4 produce a significant higher weight gain than do feeds 1 and 2.

The output appears below:

```
Grouping variable is GROUP
Analysis variable is OBS

Group Means and Standard Deviations
-----------------------------------
  1: mean = 61.025          s.d. = 5.0822              n =  4
  2: mean = 78.125          s.d. = 1.4886              n =  4
  3: mean = 89.0667         s.d. = 4.4276              n =  3
  4: mean = 85.775          s.d. = 3.5976              n =  4

Analysis of Variance Table

Source           S.S.         DF            MS         F        ·Appx P
-----------------------------------------------------------------------
Total           1923.41       14
  Treatment      1761.24        3          587.08     39.82       <.001
  Error           162.17       11           14.74

  Error term used for comparisons = 14.74 with 11 d.f.

                                                              Critical q
Newman-Keuls Multiple Comp.         Difference   P    Q         (.05)
-----------------------------------------------------------------------
       Mean(3)-Mean(1) =              28.0417    4   13.523      4.256 *
       Mean(3)-Mean(2) =              10.9417    3    5.277       3.82 *
       Mean(3)-Mean(4) =               3.2917    2    1.587      3.113
       Mean(4)-Mean(1) =              24.75      3   12.892       3.82 *
       Mean(4)-Mean(2) =               7.65      2    3.985      3.113 *
       Mean(2)-Mean(1) =              17.1       2    8.907      3.113 *

  Homogeneous Populations, groups ranked

              Gp Gp Gp Gp
               1  2  4  3
                     ------
              ---
                  ---
```

This is a graphical representation of the Newman-Keuls multiple comparisons test. At the 0.05 significance level, the means of any two groups underscored by the same line are not significantly different.

Step 8: To display a graphical interpretation of the analysis, **click on the "Graph" button** on the View Results Window to view the graph associated with this analysis. The default graph shows a box and whiskers plot comparison as shown below:

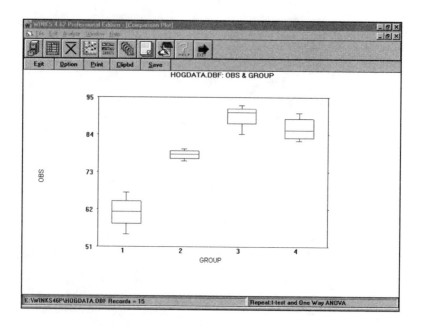

Step 9: From the group comparison plot, you can select from among a variety of display options. **Click on the "Options" button** to display the options menu for the graph.

Step 10: For this example, check the following options (and uncheck all other options) on Graph Options dialog box:

- Display Means
- Error Bar (+/- St. Error of the Mean)
- Connect Means
- Bar Graph of Means

The graph, as shown here, will be displayed. Print the graph by clicking on the "Print" button.

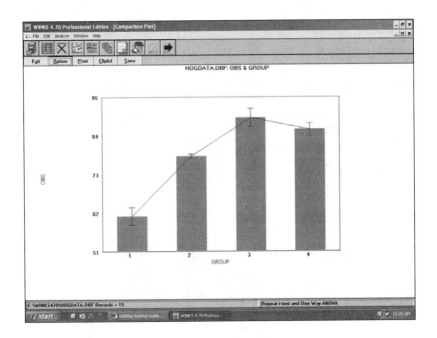

Step 11: Exit WINKS by clicking on the Exit Arrow on the WINKS icon bar or by choosing Exit from the File menu.

Tutorial 3 – Using Data from an Existing File

This example shows you how to open an existing database, display descriptive statistics on a single variable, and display a histogram. Follow these steps:

Step 1: This example uses a database placed in the WINKS folder when you installed WINKS. To open this database, **click on the File menu and select Open**. The Open dialog box appears. **Select the file named EXAMPLE.DBF**. Notice at the bottom of the screen, a message tells you that the EXAMPLE.DBF database is opened and that it contains 50 records.

Step 2: To perform a Descriptive Statistics analysis on this data click on the X-bar icon or select **Analyze, Descriptives, Detail/One Variable**. A dialog box will appear where you can choose a field name. **Click on AGE, then on Ok**, as shown here.

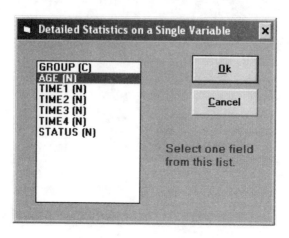

Step 3: - The output contains descriptive statistics from the data in the AGE field. Scroll this window to view information.

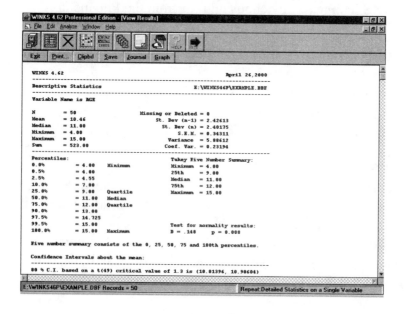

Step 4: Click the "Graph" button to display the histogram associated with this data. A screen will appear showing a histogram of the AGE data. **Click the "Bell" button** to add a bell curve (normal curve) to the graph.

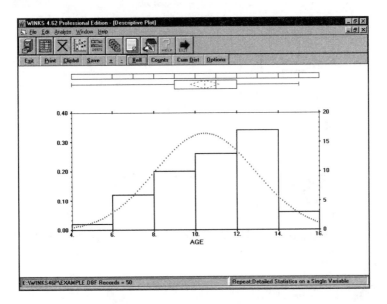

Step 5: From the Window menu at the top of the screen, **select the "Tile" option**. This causes the text and graphic windows to be displayed side by side.

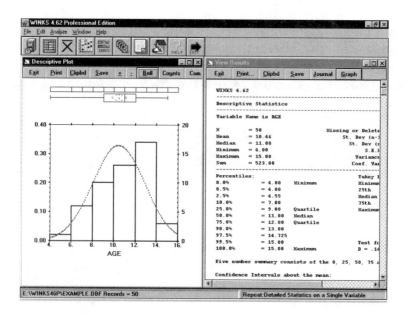

When you begin another analysis, the contents of the current
viewers are erased to make room for the information from the new
analysis.

Tutorial 4 – Using WINKS GRAPHS

WINKS uses two kinds of graphs. Most graphs, particularly those
associated with statistical procedures, are built into the WINKS
programs. A separate graphic program called First Impression is
included with WINKS to give you more graph options. This
tutorial will introduce you to both kinds of graphs.

Graph Example 1: WINKS Graphs are available though most
statistical procedures or by themselves through the
Analyze/Graphs/Charts menu item. In this example you will
display a matrix of scatterplots. This graph allows you to examine
the relationship between several pairs of variables at once. To
create this graph:

Step 1: Open the file named EXAMPLE. (File/Open Menu item)

Step 2: Select the **Analyze** then **"Graph/Charts"** and **"Correlation Matrix Graphs"** from the menu.

Step 3: From the variable selection dialog box, select the variables TIME1 to TIME4 as shown below by highlighting a variable name, then clicking **Add**.

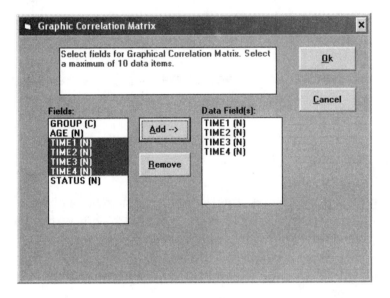

Step 4: Click **Ok** and the matrix of graphs will be created, as shown below.

Step 5: Click the button labeled **"Line"** to display regression lines on each graph.

This graph allows you to examine the relationships of several variables at once to discover which may be highly correlated and which are not.

E:\WINKS46P\EXAMPLE.DBF

Graph Example 2: *First Impression* is the name of a charting program created by Visual Tools. It is included with WINKS to give you additional graphing options. *Since it is a separate program, it contains differences in the way charts are displayed and manipulated.* Follow these steps to display an example First Impression Chart.

Step 1: Open the database BARCHART by clicking on the "File Cabinet" icon and choosing BARCHART.DBF. A message at the bottom of the screen tells you the database is opened.

Step 2: Click on the "Chart" Gallery Icon or select "Graph/First Impression Charts" from the Analyze menu. A dialog box appears displaying several chart types. **Click on the "Bar Chart"** icon in the Chart Gallery.

Step 3: Select the variable field to chart. **Click on VAR1 in the field box, then Add.** This places VAR1 in the "Data field(s)" box. You could choose as many as 10 fields. For this example, select only VAR1. If you accidentally choose a wrong field, remove it by clicking on the field name in the Data field(s) box, then click the Remove button.

Click on LABEL variable name in the Field box, then click on the _Label button. This selects what field will be used as the label field in the graph. Once you have selected the VAR1 field, and the LABEL field for the label, click **_Ok**. A bar chart appears as shown below.

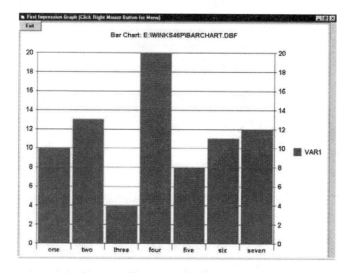

Note:

First Impression graphs have only one button bar option, Exit.
Since *First Impression* is really a separate program it contains its own menu described below.

Step 4: While pointing the mouse cursor at the plot, **click the Right Mouse Button** to display the pop-up First Impression menu and **select *Plot...*** This menu allows you to choose options on how your graph will be displayed. **Click on the 3D "radio button"** option in the "Chart" option box at the left of the dialog box, then Ok. A 3D version of the bar chart appears.

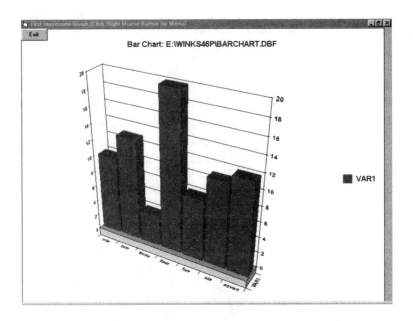

To Print a First Impression Plot **choose the Print option from the Pop-Up Menu** (right mouse button). From the Print menu, choose the "Page Setup" option to select how the chart will be printed.

Step 5: Press the right mouse button again and **select Plot**.... This time, choose a **LINE (Tape) Chart** type from the Series Type option box, then choose **Ok**. A Tape/Line graph appears. Experiment with the other graph types to see the variety of ways you can choose to display a First Impression Graph.

Notice the folder tabs at the option of the First Impression menu. When you click on one of these tabs, a different menu with charting options appears.

Tutorial 5 – Importing Data from Excel

WINKS uses a procedure in the File/Utilities menu called "Excel/Paste Import" to import data from Excel. This feature will actually work from any Windows spreadsheet or database program that supports the Windows clipboard.

The following is an example of how to use this procedure. To perform this tutorial, you must have Microsoft Excel installed on your computer.

Step 1. Begin Excel. Open the Excel file named EXAMPLE.XLS in the WINKS directory (usually in the folder C:\WINKS470.)

Step 2: In Excel, highlight the range of cells you want to import (in this case from A1 to G51) as shown below. **Choose Edit/Copy or click on the Excel Copy icon.**

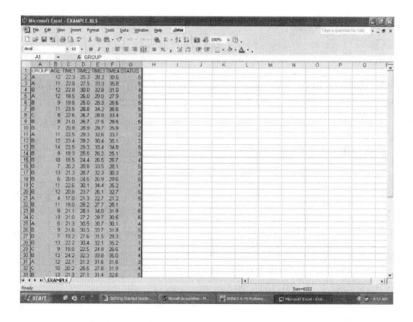

Notice in this selection that the first row contains the names of the column variables. When performing an Excel Import:

- You can copy cells that contain only the data values. Or,

- You can have the top row of cells contain column labels that will be translated by WINKS into variable names.
- We recommend that you use the second option as it makes the import easier.

Step 3: Switch to the WINKS program. From the WINKS menu **select the File/Utilities "Excel/Paste Import Option."**

Step 4: You will be asked if the top row of cells contains variable names. In this case, **click Yes.**

Step 5: WINKS imports the spreadsheet into the program. **Select File and Save** to save the data in a WINKS file so you can use it for analysis. (See the Excel tutorial on www.texasoft.com for more info.)

Tutorial 6 – Pasting output into Word

Step 1: Redo Tutorial 3 up to Step 3 where you see the output in the WINKS viewer.

Step 2: Click on the **Clipbd** button to place the contents of the viewer into the Windows clipboard.

Step 3: Open Word (or any word processor) and click on Paste.

Step 4: Highlight the pasted output and change the font to Courier New. This last step is important to make the output tables line up correctly. Always change WINKS output to Courier New in your word processor for best results.

Note: Clipbd and paste also works with WINKS graphs.

Part 2 – Data Topics

WINKS Data Topics

- Creating a New Database
- Do You need a Grouping Variable
- Database Limitations
- Data Examples
- Main File and Edit Menus
- Spreadsheet File and Edit Menus
- Data Editing and Manipulation
- Entering and Importing Data
- Printing a Data Report

This chapter explains how to create, enter and manipulate data within WINKS and how to create a data report.

The process for entering data and performing an analysis can be summarized in three steps:

Step 1: Create a database.
Step 2: Enter data into the database.
Step 3: Choose an analysis option.

The following sections explain how to create and enter data into WINKS.

Creating a New Database

Before you can enter data into a database, you must create a new database. The "New Database" option on the File menu is used to create a new database. The structure, or layout, of a database must be described before you enter your data. WINKS allows you to create a new database in two ways:

- Choose from a predefined structure or
- Create a customized database structure.

Creating a Database from a Pre-defined Structure

A pre-defined structure is a "blank" database designed for a particular analysis type. Using a pre-defined database allows you to enter data for your analysis quickly. The list below contains examples pre-defined databases in WINKS. For example, if you're performing an independent group t-test, choose the option "Independent Group t-test or ANOVA." The proper database will be created and then you can enter your data into the database. Here are some of the pre-defined database options:

- Create a customized database
- Single variable, Detailed Stat, Histogram, Stem & Leaf
- Time Series Lines with Grouping Variable
- Statistics From Count Data, VALUE, COUNT
- Grouped Histogram, Statistics or Stem & Leaf, 1 Value, 1 group
- Pie Chart: Label and Value
- Bar Chart, Group and Two values
- Bar Chart, Group and Up to 5 values
- X-Y Chart/Scatterplot, 2 Values
- X-Y Chart/Scatterplot with GROUP, 2 Values, 1 GROUP
- By Group Plot, One Group
- By Group Plot, Two Groups
- Independent group t-test or ANOVA
- Paired t-test or McNemar's Test
- etc...

For an **example of creating a new database using a predefined structure**, see Tutorial 2 in Part 1 of this manual.

Creating a Customized Database

If none of the pre-defined database structures meet your needs, you need to create a customized database. When you choose to create a customized database, you need to consider these items:

- What fields are required for the analysis?

- Do you need a Grouping Variable?

The following sections describe how you create a customized database structure to match your analysis needs.

Specifying Field Types Required by an Analysis

When you create a new database, you must specify certain information about each data field, including the field name, type, width and number of decimals (if any):

1. The **FIELDNAME**: A fieldname is 1 to 10 characters in length, MUST begin with a character (a to z) and can contain letters, numbers and the underscore character "_". Upper and lower case DO NOT matter, since the name is translated into all upper case.

2. The **TYPE**: Type may be . . .

CHARACTER - May contain any characters.

NUMERIC - Must contain numbers only. Example permitted numbers are:1.00, -4.32, 6, 10000. Example illegal numbers are: 450-23-1232, $23.95, 123 Maple. (For data like these, use the character type.)

DATE or LOGICAL fields can be created in WINKS, but WINKS analyses only uses numeric and character fields.

3. The **WIDTH** of the field: Choose a width so that the maximum number of characters needed will fit into the field. Watch out for negative numbers, a "-2" takes up two characters. Also, count the decimal point as one character. For DATE or LOGICAL types, field widths are automatically set at 8 or 1 respectively.

4. DECIMALS: Decimals are only valid for numeric fields. This specifies to WINKS how many decimal places to retain in the field. For example, if you wish to store numbers that are dollar prices, your data may look like "9999.99". This field would have a width of 7, with 2 decimals.

Customized Database Example

This example shows you how to enter a customized database. The data is listed below. Each row is one student. GRADE is the grade received in the class, age is age, SEX is gender, WT is weight and SCORE is the score on a 25 point pre-test.

Step 1: From the *File* menu, choose "New database." Enter the name MYDATA (or a name of your choice, limited to 8 characters).

Step 2: A list of database structures will be displayed. For this example, choose the "Create a customized database" option.

Data for the customized database example:

OBS	GRADE	AGE	SEX	WT	SCORE
1	A	18	M	165	22.3
2	B	19	M	145	22.8
3	B	17	F	122	22.8
4	C	18	M	196	18.5
5	B	17	M	188	19.5
6	B	18	F	140	23.5
7	C	19	F	121	22.6
8	B	20	F	112	21.0
9	C	19	F	122	20.9
10	A	18	M	176	22.5
11	B	18	M	165	23.3
12	A	19	M	135	21.8
13	A	18	F	121	24.8
14	C	19	M	186	16.5
15	B	17	M	148	18.5

Step 3: Define the database structure: for each field (each column of data) in the database, specify the following information:

- A name of the field - something to identify it
- The type of data - is it numeric or character?
- The width of the field - enough to hold the biggest entry
- Decimal places - if needed

For this example, you will use the following information:

Field name	Type	Width	Dec
GRADE	C	2	
AGE	N	3	
SEX	C	2	
WT	N	4	
SCORE	N	5	1

The GRADE and SEX variables are of type "C" (character) and the rest of the variables are numbers "N". Only the score variable requires a decimal value. Enter the information about the database structure into the database definition screen. The final structure definition screen should look like the following screen.

Specifying the structure of a database

Notice that the widths defined here are actually 1 character wider than actually needed. If you are not pressed for space in the database, this will make your listings easier to read.

Step 4: A data entry screen appears listing the names of all of the fields. Enter the data into the spreadsheet.

Step 5: Before exiting the data entry screen, you can correct errors in data entry mode by moving your cursor to a cell and re-entering data. See sections later in this chapter concerning other techniques of manipulating the database. Exit and save the database by selecting the "exit" option from the *File* menu. You will return to the WINKS main menu screen.

Step 6: Once you have entered your data into the database, you are ready to perform one or more analyses. For example, from the *Analyze* menu choose the "detailed statistics" option. See Chapter 4 for more information about using the *Analyze* menu options.

Do you need a Grouping Variable?

A grouping variable is a code that tells WINKS how to separate data values into groups -- such as control group versus experimental group. If your data contains information on two or more groups, you should include a grouping variable in the database. A grouping variable could be character (e.g., a,b,c etc.) Or numeric (e.g., 1,2,3, etc.).

For example, suppose you will be comparing the mean heights of two groups. You could choose a grouping code to be 0 and 1, 1 and 2, a and b, old and new, or any other designation that makes sense to you. You can then visualize the database as consisting of two columns of information, group and height:

```
GROUP      HEIGHT
1          60.4
2          55.9
1          60.2
1          58.0
2          54.3
etc...
```

From this example you can see how WINKS can now tell that the height 60.4 belongs to an individual in group 1, the height 55.9 belongs to group 2 and so on.

WINKS Database Limitations

- Maximum of 250 fields.

- Maximum width of a field name is 10 characters.

- Maximum width of a cell is 60 characters (15 for numbers).

- Dates are always 8 characters and logical fields are 1 character wide.

- Memo fields are not supported.

- Maximum of 32,000 records

Note: See Help/Print Supplementary Docs/Limits for more information on specific limits on each analysis type.

File Menu

The *File* and *Edit* menus on the main WINKS menu screen are used for the following purposes: (see also the Spreadsheet **File** and *Edit* menus described later in this chapter.)

New database
Create a new database (data set for analysis). See the tutorials in this chapter for examples.

Open a database
Opens an existing WINKS (or dbf) database. Use this option to choose the database that you will be analyzing.

Once a database is open, you will see its name at the bottom left of the screen, along with the number of records in the database.

Delete a database
Deletes a database and its related missing values files (if any.)

Utilities - export, import, report, sort
This menu provides several data utilities including:

- "Export Data" -- This option allows you to write out the data in a database to an ASCII file. This option is useful when you want to transfer data from your WINKS database to another program that does not read .dbf files. There is more information on exporting later in this chapter.

- "Data Report" -- Allows you to produce a report using the data in your database. See the section titled "Printing a Report" later in this chapter

- "ASCII Delimited Import" -- See the section below titled "Entering and Importing data" for information on how to import data into a WINKS database.

- "123 (WKS) File Import" - Import data from a Lotus 1-2-3 file. See the section below titled "Entering and importing data."

- "Excel/Paste - Import Excel or other spreadsheet data – see example later in this chapter.

Journal options -- The Journal option displays a submenu allowing you to clear, print, edit, or delete your output journal.

Print setup -- Allows you to choose your printer type.

Exit -- Use this option to end the WINKS program.

Edit Menu

The *Edit* menu opens the Spreadsheet data editor. If you have not opened a database, you will be prompted to select a database. See "Spreadsheet File and Edit Menus."

Spreadsheet File and Edit Menus

Once you display the spreadsheet database editor, you will notice that it also has *File* and *Edit* menus. These menus are specific to the spreadsheet, and have different options than the main *File* and *Edit* menus.

The data spreadsheet is used for entering or editing data in WINKS. To enter data into the spreadsheet, simply click on a cell and enter a value. When you press enter, the spreadsheet moves the cursor to the next cell, allowing you to enter another value. The following options are available in the WINKS spreadsheet:

- The File/"Save" option saves the database under the current database name.

- The File/"Save as" option saves the database, and allows you to specify the filename.

- The File/"Display structure" option allows you to see a list of all fields in the database, and their attributes.

- The File/"Zap" option allows you to quickly erase all records from a database but retain the database structure. To use this option, edit a database, then choose File/"Zap". You can save the blank database, or import data into the blank database and save it under a new name.

- File/"Append from ASCII file" or "Append from dBase file" option allows you to add new records into the database currently being edited. You may append data either from an ASCII file or dBase file. See more import information later in this chapter.

- File/Print Data prints a copy of the spreadsheet to the printer.

- The File/"Exit without saving" option allows you to exit the spreadsheet, and abandon all changes you have made.

- The File/"Exit" option exits the spreadsheet. You will be asked if you want to save any changes before exiting.

Options on the Edit menu are:

- The Edit/"Missing values codes" option allows you to enter a missing value code for all fields. These codes tell WINKS that data is missing from a particular cell in your database, and the information is used when performing an analysis.

- The Edit/"Pack data" option permanently erases all records (rows) marked for deletion (see mark and unmark below) you must save the database for the pack to take effect using "save as" or "save."

- Edit/"Copy" copies selected information into the clipboard.

- Edit/"Cut" copies selected information into the clipboard and deletes the information from the spreadsheet cells.

- Edit/"Paste" places information from the clipboard into the spreadsheet. This is a way to move data from another spreadsheet such as excel into the WINKS spreadsheet. However, you must make sure the WINKS spreadsheet has enough rows and columns to contain the information you paste.

- The Edit/"Column(s) and Row(s) delete" options allow you to delete columns or rows. First highlight the column(s) or row(s) to delete, then select this option. Removing or inserting rows causes WINKS to remove any formulas associated with the spreadsheet. See "Remove formulas" below.

- The Edit/"Column(s) and Row(s) insert" options allow you to insert blank columns or rows into your spreadsheet. Place your cursor where you want the insert to take place, and choose one of these options. Removing or inserting rows causes WINKS to remove any formulas associated with the spreadsheet. See "Remove formulas" below.

- The Edit/"Modify column attribute" options allows you to specify or change the field name, width, type, number decimals and missing values code used for the column/field. You can also change these items by double clicking on the fieldname button at the top of the column.

- The Edit/"Formula entry" option allows you to enter a formula to be used in an entire column or in a highlighted area. For example, you might enter a formula such as height/age. This would cause the selected cells to display the results of that formula. WINKS will remember the formula, and will automatically perform the calculation whenever you enter a new number in the spreadsheet. See *Help* for more information.

- The Edit/"Remove formulas" option is used if you do not want WINKS to remember formulas you have entered.

When you remove formulas, the numbers in the effected cells remain, but are no longer updated when you enter new numbers elsewhere in the spreadsheet.

- The Edit/Recode option allows you to recode data values in the spreadsheet.

- The Edit/"Text replace by column" option allows you to replace the contents of text fields (cells) with a specified text string.

- The Edit/"Unmark/Mark records" option allows you to mark or unmark records for deletion (see pack database). When you mark a record for delete, an asterisk "*" appears next to the row number. Marked records are ignored during analysis, so this is a good way to eliminate some records temporarily from an analysis. Marked records are not deleted from the database unless you perform a pack.

- The Edit/"Mark or Unmark records by example" option allows you to mark or unmark records that meet a specified pattern. For example, if you want to mark all records for which the field sex is m, place your cursor on a cell containing an m, then select this option. This is a quick way to perform an analysis on a subset of your data.

- The Edit/"Unmark all" option unmarks any records that are currently marked.

- The Edit/"Reverse marks" option reverses all marks, so unmarked records become marked, and marked records become unmarked.

Items on the Options menu are:

- Options/Ascending sort -- allows you to sort the spreadsheet on a single variable from lowest to highest

- Options/Descending sort -- allows you to sort the spreadsheet data on a single variable from highest to lowest

- Options/Random sample -- allows you to select a random sample of any size from the current data set.

Data Editing Techniques

The first part of this chapter introduced you to the database spreadsheet options available in WINKS. The remainder of the chapter gives details concerning several topics concerning data entry and manipulation.

Missing Values Codes

Sometimes in the collection of data there are values that are lost or cannot be gathered. These are called "missing values". When such values occur, it is important for the program to know that the values are missing so that statistical calculations may take this into account. Missing values are usually designated as an impossible value. For example, the missing values designated for the variable AGE may be -9, since it is impossible for the variable AGE to have the value -9. When the program is asked to calculate the mean of AGE, for example, it will ignore those records where AGE is -9 in that calculation if -9 has been specified as the missing value code. In most WINKS procedures, there is a casewise deletion of the record from calculation whenever a missing value is encountered.

Once you designate a missing value code for a variable, it is up to you to make sure that this code gets placed into your database in the proper records and fields. For example, if you have designated -9 as the missing value code for AGE, you must make sure that in your database a -9 appears in the field AGE if that data is missing or unknown.

A standard dBase III file does not have a way to designate missing values, but WINKS allows a way for you to designate these values in this program. The "indicate missing value codes" option on the *File* menu is used to set up these values. When this option is selected, the program will display an entry screen that is similar to a data entry screen. You may enter one missing value for each field name. The missing value must obey the definition of the field in terms of length and type.

Once missing values are entered, they are stored on disk in a file named filename. MV, where "filename" is the name of the designated database. If a new variable is created using the transformation procedure, its missing value is appended to the missing value file.

You may change or correct the missing values for a database at any time by calling up this option. If missing values are already designated for the database, they will be displayed on the entry screen, and you may edit them or accept them as they are.

Note: If missing values are not used, and there is a blank numeric variable in a calculation, it may be treated like the value 0 (zero), so it is important to use missing values if your data contains such entries. Otherwise, the statistical calculations may be in error!!

Sorting the Database in the Data Editor
While editing a database in the data editor, you can select a column, then choose either "Ascending Sort" or "Descending Sort" from the Options menu. This will sort your database according to the data in the selected column. If you wish, you can then save the sorted database. The old sort in the *File* menu has been retained for those who prefer its use.

Selecting Random Samples
To select a random sample of records from a database:

   1. Open a database then select edit.

   2. In the data editor, select Options/Random Sample.

   3. Enter the number of records you want in your sample.

The editor will randomly mark records until there is only the number unmarked that you want to use. Once the random select is made, you have two choices:

   1. You can select Edit/Pack to permanently delete all of the marked records, so you will only have the sample remaining in the database.

   2. Save and use the database with the marked records still in the database. Any analysis you perform will ignore those marked records. You can reedit the database later and select another random sample.

For instance, open the EXAMPLE database, then select edit. From the Options menu, select Random sample, then choose 20. You will notice that 20 records will remain unmarked, and 30 will be marked.

Select File/Save As and save the database under the name SAMPLE. Then select Analyze/Detailed statistics and request statistics on the AGE variable. Notice that the "N" for the statistics is 20.

Return to Edit, and select another random sample, of size 15. Then exit and save (again to SAMPLE) and repeat the Detailed analysis. This time only the sampled 15 records are used. You can return again to the editor and use the Edit/Unmark all option to unmark all records from the database, then save it. Using this procedure, you can take random sample from a database without actually getting rid of any records.

Formula Entry and Text Replacement

Use mathematical expressions to calculate new numbers from existing numbers in the database fields. For the latest information about mathematical expressions available, view the supplemental documentation file named formula.doc or refer to help. The following information shows how the "formula entry" option can be used to calculate new values. The standard arithmetic operators available are:

```
ADD                  +        SUBTRACT          -
DIVIDE               /        MULTIPLY          *
EXPONENTIATION       ^
```

Following are a few examples of correct expressions:

```
AGE/HEIGHT                    VAL1-VAL2
(AGE*TIME1)+3.2               (PI*RADIUS)^2
```

Several functions are also available such as LN, LOG, SQRT, SIN, ABS, INT and more. See the supplemental documentation or help for a complete list of functions. Example use of functions are:

```
LN(Age)                       SQRT(AGE*HEIGHT)
SIN(SCORE)                    INT(NUMBER)
```

The "Text replacement" option on the *File* menu is another way to place information into cells. You can highlight an entire row, and place text in all cells, or you can highlight one or more cells, and replace text in only those cells using the "text replacement" technique.

Note: If you have problems calculating a value, try using parentheses or dummy multiplications. For example, instead of

```
2*var1+2*var2)
```

use

```
(2*var1)+ (2*var2)
```

Data Editor Recode

The recode option in the WINKS editor makes it simple to restructure the information in your database. To recode data you must have a numeric column (field) ready to receive the data. You may need to use the Edit/Insert Column option to create a new field before performing the recode. To perform the recode in the data editor:

1. Select the "Recode" option from the Edit menu.

2. The recode dialog box is shown in figure 3.

3. Select the field name where the **results** will be placed. Also select the field used for the **recode comparison**.

4. Fill in the recode comparisons and results fields. Note that the comparison fields are left inclusive. That is, if you specify a comparison of between 1 and 2, the program will check for: "If field is greater than or equal to the value of the left field and less than the value of right field." For a recode of a single number, put the same number in both fields.

5 If applicable, select what value to recode if the comparison field contains a missing value.

6. Click OK. Check your results.

Entering and Importing Data

To enter data from a file into a new database, use the "import" options on the main WINKS File/Utilities menu. If you already have a database created, you can append (or add) new records to that database using the "Append" options on the Data Editor *File* menu.

To append records to an existing database, first open that database and go to the Data Editor. When you choose the "Append records..." option from the Spreadsheet *File* menu, you can select to append from an ASCII file or from a database (.dbf) file.

Entering Data From an ASCII File

When you choose to enter data from an ASCII file, you will be asked to specify the name of the input file.

The data from the ASCII file will be entered into the database. If there are already records in the spreadsheet, the new data from the ASCII file will be appended (added) as new records to the spreadsheet. Note: data will be entered according to the widths of each field in the database. If the data does not match the fields, the imported data will not properly be displayed in the fields.

Appending Data from Another dBase File

To append data from another dBase file into an existing database, first open the dBase file and go to the Data Editor. Then, select the option to Append data from a dBase file. Note: the data in the file to be appended must contain the same field names as the spreadsheet currently being edited.

Moving Data from Another Program

Some programs, such as Lotus 1-2-3, Excel and others can save data as a dBase file. Look in your manual's index for dBase or dbf to see if your program will produce a dBase file. If it will, this is usually the easiest way to move data from another program into WINKS.

If you have a Lotus 1-2-3 wks or wk1 file and cannot use the Lotus program to translate it, you can use the WINKS import feature from the "Utilities/Import" option from the *File* menu on the main WINKS menu screen.

This translation facility will translate 1-2-3 version 1a files. Symphony and other versions of wk* files may or may not be able to be translated with this program. However, it seems to work okay on lotus 1-2-3 2.1 and twin "wkt" files and most other wk* files. This translator will not work with Lotus version 3.x or higher files.

The program will ask for a range of the table to translate. This must be in the form such as a1.r5, where a1 is the top left of a rectangular region, and r5 is the bottom right. You will also be allowed to pick the type of output format for the data, either an ASCII or dbf file. To allow the data to be used in WINKS, choose to convert to the dbf format. An example file on disk to translate is test.wks, which contains data in cells a1.h6.

You are given a choice to let the program pick the names of the resulting dBase fields, or to specify them yourself. By default, the program will choose names such as columna, columnb, etc. If you choose, you can specify your own names, and you will also be able to specify certain field characteristics such as type (program assumes character, you may specify character or numeric) width (program assumes width from wks file), and in the case of a numeric field, number of decimals.

Excel Spreadsheet Import

To facilitate moving data from other Windows programs, WINKS uses the procedure in File/Utilities menu called Excel/Paste Import. It allows you to import data by copy and paste. Although the name "Excel" is used in this feature, it will actually work from any Windows spreadsheet or database program that supports the Windows clipboard. To use this procedure:

Step 1: Highlight and copy a range of cells in Excel you want to move into WINKS. You have two ways you can do this.

- You can copy cells that contain only the data values. Or,

- You can have the top row of cells contain
  column labels that will be translated by
  WINKS into variable names.

Step 2: Begin or switch to WINKS, and select the
File/Utilities "Excel/Paste Import Option."

Step 3: You will be asked if the top row of cells contains
variable names. If the answer is yes, click on "Yes", if not,
click on "No"

Step 4: WINKS will import the spreadsheet into the
program. If the top row of cells did not include column
names, you can double-click on each column label and
enter a new name for that variable before saving the
spreadsheet. (See tutorial in Part I.)

Importing Nondelimited ASCII files

If the program you are using does not support dbf or worksheet
files, it probably allows you to output data as ASCII files. These
are also called "text" files, "DOS" files, or "sdf" (standard data
format) files. To move data into WINKS using an ASCII file,
make sure the file is of the form described above in the section
titled "entering data from an ASCII file." sometimes you may have
to edit the text file output from a program using a text editor to
make sure the columns of data are all in fixed columns. You can do
this with an editor such as WordPerfect. Make sure you save the
resulting file in ASCII (dos, text) format. Then, follow the steps
outlined above to import the ASCII data into a WINKS database
file.

Importing Comma Delimited ASCII Files

If your program outputs comma delimited ASCII files, that is,
there is a comma between each field, WINKS can import this data
using the "comma delimited" option in the file utilities file menu
option on the main WINKS menu screen.

The data to be imported can contain number and character fields.
Character fields must be enclosed in quotes "". An example file on
disk is excomma.dat. The first few lines of this file are:

```
"a",12,22.3,25.3,28.2,30.6,5,"text"
"a",11,22.8,27.5,33.3,35.8,5,"text"
"b",12,22.8,30.0,32.8,31.0,4,"text"
"a",12,18.5,26.0,29.0,27.9,5,"text"
```

The import procedure looks at the first line of the file to determine how many fields to create. This file has 8 fields. The first and last are character. The fields will be named var1, var2, etc. You can change these names in the "Modify database" option, main menu. The import will attempt to create widths that will allow full storage of numbers and text. Everything after that is automated.

Printing a Data Report

You may output a listing of the data in the dataset (or a selected subset of the database) by using the report facility. To use this procedure, choose "Data Report" from the *File/Utilities* menu in the main WINKS screen. In this procedure you may specify the following report features:

- Which data fields to output
- Output record number as a column
- Title
- Number of lines per page
- Output a subset of the data (search)

Note: You may want to place a coded variable in your data set which will allow you to easily select a subset of data to output. To select a subset, uncheck the "Output all Records to Report" option. Subset searches can be:

1) Exact: Case is ignored.

2) First one or more letters in a field: (al* matches allen, albert, etc)

3) Keyword: match a letter pattern within a field (i.e., [al] matches allen, bales, etc).

Report results will be displayed in the viewer, where you can choose to print or save the report for future use.

If the report is too wide to fit on a single width of the specified paper width, the report will be printed in parts. A partial report from the example database is displayed below:

EXAMPLE REPORT

RECORD	GROUP	AGE	TIME1	TIME2	TIME3	TIME4	STATUS
1	A	12	22.3	25.3	28.2	30.6	5
2	A	11	22.8	27.5	33.3	35.8	5
3	B	12	22.8	30.0	32.8	31.0	4
4	A	12	18.5	26.0	29.0	27.9	5
5	B	9	19.5	25.0	25.3	26.6	5
6	B	11	23.5	28.8	34.2	35.6	5
7	C	8	22.6	26.7	28.0	33.4	3
8	B	8	21.0	26.7	27.5	29.5	5
9	B	7	20.9	28.9	29.7	25.9	2
10	A	11	22.5	29.3	32.6	33.7	2

It is highly recommended that you output a report of your dataset, to use in proofing your database entries, before performing analyses on the data.

Part 3 – Statistical Concepts Review

Using Statistics to Analyze Information

This chapter discusses some of the statistical concepts used in WINKS. If you are a little rusty on statistical nomenclature or on how to interpret the results of statistical tests, this chapter will provide a review of these concepts. If you are familiar with statistical concepts and tests, you may skip most of this chapter without missing any vital information about using WINKS. Here is an outline of what this chapter contains :

- Using Statistics to Analyze Information
- Summarizing Information with Statistics-Quantitative and Qualitative data
- Investigating Associations Between Variables - Regression, Correlation and Crosstabulations
- Using Statistics to Make Comparisons
- Performing a Statistical Test - Hypotheses, p-values and multiple comparisons
- Choosing the Right Procedure to Use

Today's world is filled with information. The computer has enabled us to gather and create more information than you can

possibly remember and understand. Computers contain
information such as company sales, bank balances, opinions on
products -- an almost innumerable list of numbers and figures.
What can you do with it all?

Usually, you do not want to look at the "raw numbers" that have
been collected. There is simply too much to comprehend. What
you want is a summary of the information. You want to reduce
thousands of numbers into a few explanatory numbers that will
give you an idea of what is going on. For example, you could look
at the daily sales figures of Mary and William's Lemonade stand
(365 numbers), and get some idea of the range of sales, but
wouldn't you rather just have a few numbers such as:

Total yearly sales: $12,521
Average monthly: $1,043.42
Lowest month: $543.04
Highest month: $1623.21

Perhaps Mary and William actually had two stands. One day they
operated on the corner of PENN and BRYAN and on other days
they operated at the corner of MAIN and BROAD. They want to
know which location is better. Again, you could look at the raw
numbers, but you probably would rather know:

Average weekly at PENN and BRYAN was $302.32
Average weekly at MAIN and BROAD was $178.29

Now you have some evidence to make a decision about which
place was better.

These two examples illustrate two major aspects of statistical
analysis - description and comparison. Another aspect of statistical
analysis that is often used is examining the association between
variables. For example, you might be interested in the relationship
between the high temperature for the day and the amount of sales.
You might suspect that the hotter the temperature, the more
lemonade sales. Would you rather investigate this by looking at
365 temperatures and 365 sales figures, or would you rather be
able to look at one or two numbers that would confirm whether or
not this relationship exists? A measure of the strength of the linear

relationship between two independent variables is called correlation. The procedure that allows you to predict the value of a dependent variable given one or more independent variables is regression.

Usually, it is impossible to gather responses from the entire population under investigation. For example, you may wish to investigate the relationship between temperature and sales for all lemonade stands in the city one summer, but do not have the time to keep records on all of them. In such a situation, a random sample is taken and statistical analyses are done on the sample in order to test certain hypotheses about the population. That is, you might randomly choose a few of the stands and analyze their records.

Three ways in which statistics are used to analyze information are description, comparison and association. The procedures in WINKS allow you to summarize information, display it graphically and perform these kinds of analyses on your data. The following sections describe the process of performing a statistical analysis and interpreting your results. Further explanation and examples are found in Chapter 4.

Summarizing Information with Statistics

Information comes in a variety of forms. For example, you may have a list of heights of all boys in a PE class. You may also have a count of how many have black hair, how many blond, how many brown and how many red.

There are two very different kinds of information. The first type is often called *quantitative* or measurement data, since the numbers used in measuring height measure a quantity (where averaging makes sense). The hair color data are often called *qualitative* data - - color names some quality, but it has no rank or order. Black hair does not come before red hair, etc. Although there are finer ways of describing data, quantitative and qualitative will suffice for this discussion.

Describing Quantitative Data

You often hear information from the news media such as . . . the average miles per gallon for a Ford is Z . . . the average height of 10 year old girls is W inches, and so on. These statistics are all descriptive. They give us an idea of the magnitude of some measure of location - often called the central tendency of the distribution (i.e., a collection of observations of quantitative values of interest.) The arithmetic average is often used as the measure of central tendency, although there are other measures of central tendency that can be used, such as the median or mode.

The measure of central tendency does not give the whole picture. For example, you know that if a reporter says that the average rainfall for June is 5.25 inches that this number is an average, and that the actual rainfall for June is likely to be lower or higher than this average. Suppose you were also told that the rainfall is usually somewhere between 3 inches and 8 inches. This range of likely rainfalls is called a measure of dispersion. Dispersion gives us some indication as to how close you might expect an occurrence (rainfall in June) to fall to its measure of central tendency. If the weatherman tells us that in most years, June rainfall is between 3 and 8 inches, then you would tend to believe that a year with a 12 inch rainfall in June would be a rare event -- but it could happen.

The mean: When using statistics to describe a collection of quantitative observations, you often use the mean as the measure of central tendency and the standard deviation as a measure of dispersion. The mean is also known as the arithmetic average. Thus, if you have 4 numbers:

5, 3, 4 and 4

the mean is calculated by adding up the numbers to get 16, then dividing by 4. The mean for this group of numbers is 4.

The median: Another commonly used measure of central tendency is the median. This number is calculated as the "middle" number in a list of numbers, when the numbers are ranked from smallest to largest. Thus, if you rank our current data, you would get

3, 4, 4, 5

Since there are an even number of data points, the median is the average of the two middle numbers. In this case, the median is

(4 + 4) / 2

or 4 -- which happens to be the same as the mean, but the median is NOT always the same as the mean.

For example, the set of numbers

2,2,3,4,4,5,29

has mean 7 and median 4 (the middle number). Notice that the medians of these two data sets are the same, 4, but the means are different. As illustrated by the second set, the mean is more susceptible to extreme values. Sometimes the median more accurately describes the majority of the data.

The range: In addition to a measure of location, another statistic is often used as a measure of dispersion (to indicate the spread of the data.) There are several measures of dispersion to choose from. A very commonly used statistic is the range of the data. The range is the largest number minus the smallest number. For example, in the first data set above, the range is 5 minus 3, or 2. In the second data set, it is 29 minus 2, or 27.

If you know that the mean of the data is 4 and that the range is 2, then you know that the data are "tight" around the mean. However, suppose you were told that a river bed had an average depth of 3.5 feet. Would you wade across? Maybe. Then, what if you were told that the range of depths was 14 feet? That would mean that somewhere in the river, there was a spot well over your head. Now would you walk across? You can see that a measure of dispersion is important in examining the distribution of a set of data.

The Standard Deviation: Another very commonly used measure of dispersion is the standard deviation, which measures the average difference (deviation) of the numbers on a list from their mean. The standard deviation is especially descriptive for a normal distribution, discussed later.

Percentiles and Box and Whiskers Plot: Another statistic that is used to help understand the distribution of the data is the percentile. The median is called the 50th percentile, which means that 50% of the data are below the median and 50% are above the median. In the same way the 25th percentile is that number where 25% of the data are below that number and 75% are above. The 75th percentile is similar. Fifty percent (75-25) of the data lie between the 25th and 75th percentiles. Suppose you have the following numbers (already ranked):

1,3,5,5,6,6,6,7,7,7,8,8,8,9,9,9,11,11,12,13,16

and you know the five percentiles described above are:

1, 6, 8, 11, and 16

then you know that about 50% of the data fall between 6 and 11, and that the range is 16 - 1 = 15 and that the middle of the data (median) is 8. These five numbers are approximately what make up the Tukey five number summary (See Hoaglin, Mosteller,

Tukey, 1983 for actual formulas.) You can draw a picture of this information by using a box and whiskers plot as illustrated below:

The left end (whisker) of this plot represents the bottom fourth of the data, the box represents the middle 50%, and the right whisker represents the top fourth of the data. The + locates the median. The median does not have to fall in the center of the box. If the data are skewed (which means that the data are clumped somewhere other than in the middle of the range) then the median (and the box) may be off center. For example, a box plot that looks like the following:

would indicate a distribution where most of the data are clumped together at the low end of the range. This tells us that there are a lot of low numbers, and a few high numbers. Sometimes there are numbers that fall much higher or lower than would be expected. These are called outliers, and are not included in the whiskers, but appear as individual points in the plot. The plot below contains some outliers:

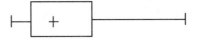

Histogram: Another graphical representation of the data is a histogram. A histogram is a bar chart in which the data are organized into groups (intervals of the continuous possible outcomes).

For example the data above could be divided into 6 intervals:

Interval	Values in Interval	Number of Values
0-3	1	1
3-6	3,5,5	3
6-9	6,6,6,7,7,7,8,8,8	9
9-12	9,9,9,11,11	5
12-15	12,13	2
15-18	16	1

(Interval includes lower, but not upper, boundary)

To make a histogram, a bar is drawn for each interval with the heights of the bars representing the number of data values (or frequency) that fall into the intervals. Comparing the bars to one another gives an idea of the proportion of total observations in each interval and thus gives an overall view of the distribution of the data.

The shape of the distribution becomes important when you are selecting the kind of statistical analysis to use. For example, many statistical procedures (parametric procedures) expect the data to have a near normal distribution, such as illustrated by the first box plot above. If the data are far from normal, you might need to use other kinds of statistical procedures (non-parametric procedures).

The Normal Distribution: A normal distribution has a graphical representation shown below. A distribution curve, such as this, is continuous with the area under the curve equal to one. The higher the curve in a given area, the more likely it is that the x's in that area will occur in that area. Most people are familiar with the concept of the bell shaped curve. Notice that it is symmetrical, with the mean located at the center of the curve.

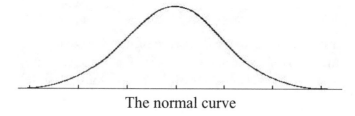

The normal curve

The bell shaped curve illustrates the distribution of a normal population. Most values are clumped at the middle of the range, with the rest trailing off into symmetric tails to both sides. The exact shape of the curve depends on the mean and standard deviation of the distribution. The mean tells where the peak (center) of the curve is located (measure of location or central tendency) and the standard deviation tells how spread out the curve is.

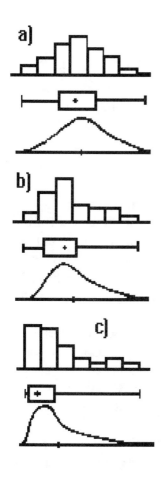

A normal distribution with a small standard deviation will be more peaked and one with a large standard deviation will be flatter. A standard normal distribution has mean 0 and standard deviation 1.

Using Histograms to Examine Data

The figure to the left shows a box-and-whiskers plot, a histogram and a distribution curve for three different distributions labeled a, b, and c. The top set of data (a) are near normal. The box-and-whiskers plot has about equal tails and the histogram shows most data in the middle with data trailing off symmetrically in both tails.

The next two distributions (b and c) illustrate various levels of skewness, and how the box- and-whiskers plot and histogram will look. Usually, you can't know the exact distribution of the data being investigated, but using the WINKS descriptive statistics procedure and histogram, you can get a good idea of the distribution and make decisions about what kind of statistic would be appropriate to use to describe the data or to perform a statistical test.

When the data are symmetrical (and near normal), the mean is usually chosen as the statistic to use to measure central tendency. If the data are skewed, the median is often the statistic of choice. This is because the mean is much more sensitive to extreme values than the median. A few extreme points can pull the mean well away from the main cluster of data. If there is more than one clump of data along the range, then neither of these measures may be as descriptive as desired.

Selecting a Measure of Dispersion

If the data are near normal, the measure of dispersion that is typically used is the standard deviation. This statistic is not simple to calculate (that's why you let the computer do it). The standard deviation can be used to place intervals around the mean where you would expect a certain percentage of the data to fall. For example, in a "normal" set of data, it is known that the range of points that consists of the mean plus or minus one standard deviation contains about 68% of the entire data set. The mean plus or minus two standard deviations contains about 95% of the data. Thus, if you know that the mean of a set of data is 20, and the standard deviation is 3, you can readily predict that the vast majority of the points may be expected to (about 95%) lie between 14 and 26 (20 plus and minus 2 standard deviations). For normally distributed data, reporting the mean and standard deviation (and the sample size) is usually sufficient to describe the distribution of the data.

In summary, you can look at a box-and-whiskers plot and a histogram of the data to determine if the data are well approximated by a normal distribution. If there is a question about what kind of test is appropriate, use this criteria: If the data are near normal, use the mean and standard deviation to describe its distribution and use parametric comparison procedures. If the data are non-normal, you may want to describe it with some other measure such as the median, the five-number summary, box-and-whiskers plot or a histogram and use non-parametric comparison procedures.

How do you know what test to perform when you have quantitative data? Look at the section titled "Choosing the Right Procedure to Use."

Describing Qualitative Data

Another commonly used type of data is *qualitative* data (sometimes called nominal, categorical or attribute data). These data typically name some attribute such as sex, eye color, pass/fail, yes/no and so forth. The data are usually not ranked. For example, blue eyes are not "greater" or "less" than brown eyes. Some nominal data may have rank such as socioeconomic class when divided into five groups 1,2,3,4,5. However in this case, group 2 is not twice as "rich" as group 1. The numbers 1,2,3,4,5 are simply convenient labels for groups. Thus means of this data would probably not make sense, so it is treated as qualitative, or categorical data. A single data set of qualitative or categorical data is often described in terms of frequencies. That is, the analysis describes how many observations fall into each group (category). For example, if you were collecting information on batters where L=left handed, R=right handed and E=either, and the data are:

L L R E L R R R R E R R R L R R L E R R R R R R

then you might summarize the data with the following information:

Type Count Percent
Left-handed 5 20%
Right-Handed 17 68%
Either 3 12%

WINKS will provide the counts and percentages in the crosstabulations procedure. To graphically describe these numbers you could use a bar chart, pictograph, or a pie chart (also in the crosstabs procedure). A bar chart of this batting information from WINKS is illustrated in the figure below.

Bar Chart: D:\WINKS\WINK303.DBF

How do you know what test to perform when you have qualitative data? Look at the section titled "Choosing the Right Procedure to Use."

Investigating Associations Between Variables

The WINKS procedures used to investigate linear relationships between quantitative variables are regression and correlation. Crosstabulations, or contingency tables, are used for this purpose for qualitative variables.

Describing a Linear Relationship Between Quantitative Variables

You may be interested in examining the relationship between two quantitative variables rather than just looking at one variable independent of another. For example, going back to the lemonade stand, you might be interested in examining the relationship between temperature and amount of sales. Your data might look something like this:

Temp	Sales
80	203
83	210
90	291
78	170
64	91
99	378
etc.	etc.

You might surmise from browsing through the data that hot days bring more sales (in general). However, if the data were not as obvious, you may want to calculate a number that would summarize this information, and you would probably like to draw a picture (a scattergram) to visually look at the trend.

Pearson Correlation Coefficient: If the data for temperature and sales are quantitative and approximately normally distributed and the relationship is linear, then the statistic that would tell you the strength of the linear relationship is Pearson's r (the correlation coefficient). This statistic ranges from -1 to 1. If the number is close to 1 or -1, it means that there is a strong association or correlation between the two numbers. If the correlation coefficient is close to 0, it means that there is a weak association or no correlation. In the case of the data above, the correlation is 0.92. This is a high positive correlation and tells us that there is strong association or correlation between temperature and sales. As temperature goes up, so do sales. (A negative correlation coefficient (for example -0.87) would imply that as the value of one variable increased, the value of the other decreased.)

Spearman's Correlation: If the data you are using are not close to normal, the correlation coefficient you should use is Spearman's r_s. This coefficient does not make the assumption that the data are normally distributed. Ranks of data, rather than data values themselves, are used to calculate Spearman's r_s. Spearman's coefficient also ranges from -1 to 1, and is interpreted similarly to the Pearson's coefficient. A value of 1 results from a perfect direct correlation of the ranks of the data, and a value of -1 shows perfect inverse correlation of the ranks.

Scatterplot: The correlation coefficient is usually not sufficient to tell the whole story about the relationship between two variables. Two variables may have a high degree of association but not in a linear relationship, in which case Pearson's r is not accurate. To check for linearity, it is always recommended that you examine a plot of the data, where each point is plotted on a graph having one variable on the horizontal axis, and the other variable on the vertical axis (such as temperature by sales.) This is called a scatterplot. The figure below illustrates what that might look like for the lemonade stand data. The scatterplot visually shows you the relationship between the two variables. If there is a strong linear relationship, the scatter will be in an approximate straight line. The more widely scattered, the smaller the correlation coefficient will be. Also, the scatterplot helps you detect points that do not fit the general pattern. A point that falls in an area where there are no other points may indicate a data entry error or an "outlier" -- an unusual point not expected. For example, if Mary and William went on vacation in the summer, there might be a point where the day was very hot, but there were no sales -- this would show up as an unusual point on the scatterplot.

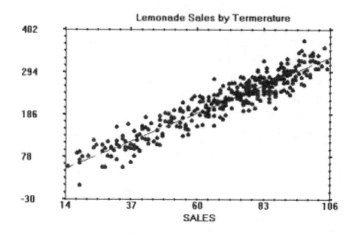

The correlation coefficient and the scatterplot combined will usually give you an indication about the strength of the linear relationship between two variables. WINKS provides a correlation calculation procedure in the Regression and Correlation procedure, which allows you to calculate a Pearson's and Spearman's correlation coefficient and to produce a scatterplot. The Descriptive Statistics and Graphs procedure also allows you to display a scatterplot.

Regression: Another procedure used to investigate the linear relationship between two quantitative variables is regression. Regression differs from correlation in that regression assumes that one of the variables is dependent on the other, while correlation assumes that both variables are independent. It may be that both are influenced by other factors and correlation will tell you whether they tend to be associated with each other without assuming that one causes the other.

Regression is useful for predicting responses for the dependent variable given the value(s) of the independent variable(s) within the range of the data, provided the necessary assumptions are met. For example, if appropriate, regression would allow you to predict lemonade sales for a given temperature.

Like correlation, regression assumes that the relationship between the variables is linear. Regression procedures also assume that the values of the independent variable (X) are fixed (without error) and that, for a fixed X value, the population of Y values (values of the dependent variable) is normally distributed and that all these normal distributions have equal variances (i.e., the variation in Y for a given X is equal for all X's). You can check these assumptions using residual (i.e., difference between observed and estimated Y) plots. If the residuals plotted against an independent variable show a pattern other than a horizontal band of points randomly scattered about zero, these assumptions may be violated.

A t-test is used in simple linear regression analysis to test for significance of the slope of the regression line, the line of least squares drawn through the scatterplot. This t-test checks whether the slope of the regression line is zero, a test equivalent to whether the correlation is significant. These tests tell you whether there is or is not a statistically significant linear relationship between the variables.

How do you know what test to perform when you are trying to find relationships between variables? Look at the section titled, "Choosing the Right Procedure to Use."

Describing a Relationship Between Qualitative Variables

If you are describing two pieces of qualitative information for each observation, you report the information with a frequency table.

For example, if you checked the players of a softball team for batting side preference and gender you might report:

	Male	Female	Total
Left Handed	3	2	5
Right Handed	10	7	17
Either	2	1	3
Total	15	10	25

A graph of this information could be created by using the WINKS
3 dimensional bar chart, as illustrated here.

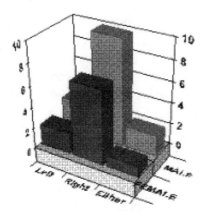

The 3-D graph allows you to visualize the relationship between the
two variables and may help you see important patterns in the data.
The WINKS Crosstabulation procedure allows you to create a
table of counts between two variables and will allow you to create
a 3-D bar chart of the resulting crosstabulation table. In addition to
describing and displaying information about data in categories, a
chi-square test procedure can be used to test whether there is a
statistically significant association (see "Using Statistics to Make
Comparisons") between the two variables or if they are
independent of each other. A chi-square test assumes that

observations are independent of one another and that each observation can be assigned to one and only one category.

For more information about analyzing this type of relationship, see the section "Choosing the Right Procedure to Use" later in the chapter.

Using Statistics to Make Comparisons

The previous sections discuss how statistics are used to summarize information into a few descriptive numbers and to examine associations between variables. Another common use of statistics is to help you make decisions about comparisons. For example, medical research is often interested in developing new ways of treating illnesses. A researcher may want to compare the effectiveness of one medicine against another. Usually, this results in an experiment where subjects are randomly assigned to two groups. For example, one group is given medicine 1 and the other is given medicine 2. Information is collected about the effect of the medicine on each individual. The purpose of the test medicine is to relieve headaches. If medicine 1 relieved headaches in an average of 33 minutes and medicine 2 relieved headaches in an average of 32.5 minutes, is there enough evidence to say that medicine 2 is "better?" Is a half a minute "significant?" Or, is the difference merely due to chance? A comparative statistical analysis is designed to allow you to answer these kinds of questions with some idea about the strength of the evidence on which your decision is based.

A variety of statistical tests are available for making comparisons of measures of location (mean or median) depending on the type of data and the number of treatment groups being compared. A single sample t-test is used for comparing a sample average to a known or hypothesized population mean. A two sample t-test is used to compare the means of two independent groups, and analysis of variance (ANOVA) is used to determine the existence of differences among the means of two or more independent groups. If the data are paired, a t-test is used (actually a single sample t-test on the average difference between observations in a pair). A repeated measures ANOVA is used for repeated measures on more than two groups. Multiple comparison procedures detail comparisons of more than two means, providing information about where differences lie.

Non-parametric procedures are available for comparing groups whose distributions cannot be assumed to be normal or where assumptions of equal variances are not met. WINKS uses the Mann-Whitney procedure for two independent groups, the Kruskal Wallis for more than two independent groups, and Friedman's test for repeated measures. These non-parametric procedures use the ranks of the data values, rather than the data values themselves, and are comparisons of medians, rather than means, of the groups. If data are dichotomous repeated measures, Cochran's Q and McNemar's tests are used to compare proportions of "successes" in the groups.

Performing a Statistical Test

As noted above and throughout this tutorial, statistical tests are used in a variety of situations to test hypotheses--about equality of means, equality of variances, significance of correlation or slope of the regression line, equality of proportions in categories (i.e., significance of association between categorical variables), equality of medians.

There is a standard method for using statistics to make decisions. All of the statistical tests in WINKS can be interpreted using the method discussed here. Generally, after checking that the appropriate assumptions are met, the steps in using a statistical test to make a decision are the following:

1. State a null hypothesis (Ho) (and usually an alternative hypothesis (Ha)).

2. Perform an analysis to test the hypothesis (the statistical test).

3. Interpret the test and make a decision, using a decision criteria based on the probability that the null hypothesis has been satisfied.

Stating the Hypotheses

A null hypothesis (sometimes called an hypothesis of no difference) usually states just the opposite of what you hope to show or suspect is true about the population parameters in question. The reason for this is that a statistical test results in a decision to reject or fail to reject the null hypothesis and it is usually considered best to make it difficult (by requiring sufficient evidence) to reject the null hypothesis. This way, accepting a change requires a significant amount of evidence. Changes or differences that may be due simply to chance rather than treatment differences are not easily accepted.

For example, in the headache medicine example the hypotheses are set up to assume that the new medicine is no better (or worse) than the old, and evidence is required to decide otherwise. To obtain this evidence or determine a lack of it, the experiment described above is done. As usual, it is impossible to test the medicines on the entire population of potential patients, so sample groups are tested. The observations on the samples are then used to test the hypotheses about the populations.

The null hypothesis might be stated as:

Ho: There is no difference between the mean times to relief in patients using medicine 1 and those using medicine 2.

The alternative hypothesis states the conclusion you will make if there is enough evidence to reject the null hypothesis. An alternative hypothesis:

Ha: There is a difference between mean time to relief for medicine 1 and medicine 2.

That is, the mean time to relief is different for the medicines.

This kind of alternative hypothesis is called <u>two-sided</u> because it allows for differences in either direction -- medicine 1 could be better OR worse than medicine 2. You could also have a <u>one-sided</u> hypothesis that the mean time to relief for medicine 2 is better (shorter) than that for medicine 1.

Performing the Analysis

How do you decide if you have evidence to reject the null hypothesis, and thus be able to show evidence for the alternative? That is, how do you use statistics to decide which of the two hypotheses to choose? The statistical test is the tool you use to make this decision.

To test an hypothesis about a population parameter (e.g., the mean), a test statistic is calculated which compares the observed data (e.g., sample average) and the expected value of the population parameter when the null hypothesis is true. If the difference between observed and expected values is large, that is, if the test statistic is extreme, it is taken as evidence to suggest that the null hypothesis is not true. If the test statistic is extreme enough, the null hypothesis is rejected and the alternative hypothesis accepted. How extreme the test statistic must be to reject the null hypothesis depends on the chosen significance level (alpha-level) of the test.

Given a significance level, and having stated an alternative hypothesis, a critical region is defined. The critical region is the range of values of the test statistic that will cause you to reject the null hypothesis. See the figure below.

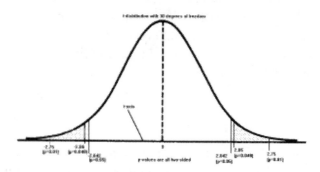

t-distribution curve

When the null hypothesis is true, the test statistic follows the distribution from which the test takes its name, for example, Student's t distribution (t-test), normal distribution (z-test), F distribution (F-test) or chi-square distribution (chi-square test). Extreme values of the test statistic (those far enough from the center to cause rejection of the null hypothesis) will fall in the tail(s) of the distribution curve. (Thus, tests are sometimes called one-tailed or two-tailed, depending on whether the alternative hypothesis is one-sided or two-sided.) The significance level determines the size of the rejection region, that is, how much of the tails are extreme enough to cause rejection. If the alternative hypothesis is one-sided and alpha is 0.05, the rejection region is the most extreme 5% of the area under the distribution curve in one tail. If the test is two-tailed, alpha is divided between the two tails and each tail's portion of the rejection region has area 0.025. Thus, if the test statistic falls in the most extreme alpha percent of the distribution (the critical region), the null hypothesis is rejected. The least extreme value of the test statistic for which rejection occurs is called the critical value.

The "t-distribution curve" above shows a Student's t distribution with 30 degrees of freedom. For a two tailed test with a significance level of 0.05, the critical values are 2.042 and -2.042 and the critical region is the area (values of t) "outside" these critical values, that is, to the right of 2.042 and to the left of -2.042.

The area of the critical region (the shaded areas combined) is 0.05, the significance level (alpha level) of the test.

A calculated t-statistic of 1.34, which falls between the critical values, is not in the critical region, and therefore does not lead to rejection of the null hypothesis. By contrast, a calculated t-statistic of 2.91 falls in the critical region and therefore indicates rejection of the null hypothesis.

If your alternative hypothesis is one-sided, care must be taken with respect to the sign of the test-statistic. The direction in which a test-statistic must be extreme in order to signal rejection of the null hypothesis depends on the alternative hypothesis.

When compared to the distribution of the test statistic under the null hypothesis, a probability of obtaining that value of the calculated test statistic (or a more extreme value) is obtained. This is called the p-value. The p-value ranges from a minimum of 0.0 to a maximum of 1.0. If the p-value associated with a test is small, there is evidence to reject the null hypothesis and accept the alternative. Many people use the value of 0.05 for the significance level to decide if they will reject the null hypothesis. That is, if the p-value for a statistical test is 0.05 or less, they reject the null hypothesis and conclude that there is evidence to support the alternative hypothesis. In WINKS, most procedures report both the value of the test statistic and the p-value. For example, you might see the following results for a t-test:

$t = 2.06$ D.F. $= 30$ p $= .048$

Interpreting the Test and Making a Decision

In the t-test reported above, the t-statistic was calculated to be 2.06. The degrees of freedom for the test (D.F., which is a value that defines the shape of the t-distribution and is related to the sample size) is 30 and the p-value is .048. A decision can be made by comparing the calculated test statistic to the critical value determined by the significance level, sample size, and hypotheses, and available from tables of the probabilities associated with the values of the distribution. A test statistic more extreme than the critical value points to rejection of the null hypothesis because it means that the observed mean (or observed difference of means) is too different from the hypothesized mean (or hypothesized difference of means.)

The 0.05 alpha-level critical value for a two-tailed t-test with 30 degrees of freedom is 2.042. That is, when the two population means are equal as hypothesized (under the null), there is a 5 in 100 chance of obtaining a t-statistic greater than 2.042 or less than -2.042. Thus, if the test statistic is greater than 2.042 or less than -2.042, the null hypothesis is rejected since there is only a small chance (less than 5 in 100 when the null hypothesis is true) that you would obtain evidence as strong or stronger against the null. In this case, the t-statistic is 2.06, greater than 2.042, so the null hypothesis is rejected.

Alternatively, a decision can be reached using the p-value. A low p-value points to rejection of the null hypothesis because it indicates how unlikely it is that a test statistic as extreme as or more extreme than the one given by this data will be observed in a sample of this size from this population if the null hypothesis is true. In this case p=0.048. This means that if the population means are equal as hypothesized (under the null), there is a 48 in 1000 chance that a more extreme test statistic will be obtained using data from this population. That is, the difference between these observed averages is so far from zero that the chance of getting differences farther from zero is less than 48 in 1000.

Is 48 in 1000 small enough? That is, does p=0.048 indicate enough evidence against the null hypothesis to reject it? It is a researcher's judgment what significance level to use. In this case, if a significance level of 0.05 is used, the p-value of 0.048 is small enough (less than 0.05) to say there is sufficient evidence against the null hypothesis. The t-statistic associated with 0.048 (2.06) is more extreme than the t-statistic associated with 0.05 (2.042).

Perhaps a more stringent criteria is necessary, say a significance level of 0.01. Then the p-value of 0.048 is too big (greater than

0.01) to call for rejection of the null hypothesis. The t-statistic associated with 0.048 is 2.06, which is less extreme than the t-statistic associated with 0.01. The critical value for a two-tailed t-test with alpha = 0.01, with 30 degrees of freedom, is 2.75.

Refer to the t-distribution graph in "t-distribution curve" figure and notice the t-values marked 2.75 and -2.75 in the tails of the distribution. The p-value associated with these t-values is p=0.01.

Suppose a significance level of 0.05 is used. You will reject the null hypothesis and conclude that there is evidence that the alternative hypothesis is true. If this had been the result of the headache medicine, then you could have rejected the null hypothesis that the medicines were equal and made a decision that the medicines had different times to relief.

Since you are comparing only two groups, you can then look at the sample means to see which is preferable, knowing the two have been found by this statistical test to be significantly different.

P-values are convenient in that you don't need a table to find the critical value to compare to the test statistic. They are useful in that they provide more than simply a reject/fail to reject decision. By comparing the p-value to the alpha level, they provide a sense of the strength of the evidence against the null hypothesis. This allows readers to know more than there is/is not strong evidence against the null. Readers can know for themselves how strong the evidence actually is and use their own judgment in making a decision.

Interpreting Multiple Comparisons

In analysis procedures that compare three or more groups, a multiple comparison test is often performed to identify pairs of groups that are statistically different from each other at a particular significance level (alpha level). There are a number of multiple comparison procedures in existence. WINKS uses a procedure called the Newman-Keuls procedure. This procedure makes pairwise comparisons of groups (usually comparing the means or mean ranks) and specifies which comparisons are statistically different at a particular alpha level (WINKS uses 0.05).

For example, in a One-Way Analysis of Variance with four groups, the test statistic (F-test) will tell you if there is evidence that the means of the four groups are different. However, this does not tell you which group means are less than or greater than other group means -- in other words, you do not know where the differences lie. In WINKS, the multiple comparison procedures produce a graph that tells you where the differences (if any) lie. Consider the following graph produced by WINKS from a Newman-Keuls multiple comparison procedure (comparing means of 4 groups):

```
Gp   Gp   Gp   Gp
 1    2    3    4
                -------
----
      ----
```

The group means are displayed in increasing order. (The mean of group 1 is smallest and that of group 4 is largest.) Any two groups underscored by the same line are not significantly different at the 0.05 level of significance. Look closely at the graph. The top line refers to the first set of groups whose means do NOT differ. In this case, the means of groups 3 and 4 are not statistically different. There are no other two group means that are NOT significantly different from each other. Therefore, all other pairs of comparisons are statistically different. Thus, you can say

- The mean for group 1 is less than the means of all other groups.

- The mean for group 2 is greater than group 1 and less than groups 3 and 4.

- The means for groups 3 and 4 do not differ from each other, but the means from 3 and 4 are both greater than the means of groups 1 and 2.

From this information, you should be able to make decisions about your experiment.

Choosing the Right Procedure to Use

The following decision tree will help you decide which analysis to perform. Start at the left of the table and work your way to the right. The terms used in the tables are:

- NORMAL - data that are from a normally distributed,whose histogram is shaped like a bell shaped curve.

- CATEGORICAL refers to count data/nominal data such as male/female, hair colors, yes/no, etc.

- INDEPENDENT refers to samples/groups that do not contain the same persons, subjects or entities.

- RELATED refers to samples where measures are taken on related entities. For example, paired data, matched samples, repeated measures.

Descriptive Statistics

Read table from left to right. ⟶

	Data Type	Procedure
When you're describing one variable	Normal	Descriptives – Mean, S,D, etc
	Not Normal	Descriptives, Median, Histogram, Stem and Leaf
	Categorical	Frequencies
	Over Time	Line Plot/Time Series
When you're describing two related variables	Normal	Pearson's Correlation
	Not Normal	Spearman's Correlation
	Categorical	Crosstbulations

Comparison Tests (t-test/ANOVA)

Read table from left to right. ⟶

Comparing a SINGLE SAMPLE to a norm	Data Type	Procedure
	Normal	Single Sample t-test
	Not Normal	Runs Test
	Categorical	Goodness-of-Fit
Comparing two groups – Samples PAIRED	Normal	Paired t-test
	Not Normal	Wilcoxon
	Categorical	McNemar
Comparing two groups – Samples INDEPENDENT	Normal	Ind. Gp. t-test
	Not Normal	Mann-Whitney
	Categorical	Chi-Square
More than two groups - REPEATED MEASURES	Normal	Rep. Measures ANOVA
	Not Normal	Friedman ANOVA
	Categorical	Cochran's Q
More than two groups – INDEPENDENT	Normal	One-Way ANOVA
	Not Normal	Kruskal-Wallis
	Categorical	Chi-Square

Relational Analyses (Correlation and Regression)

Read table from left to right. ⟶

	Data Type	Procedure
When you want to analyze the relationship between two variables.	Normal	Pearson Correlation, Simple Linear Regression
	Not Normal	Spearman Correlation
	Categorical	Contingency Coefficient
	Mixed	Spearman Correlation
To analyze the relationship between three or more variables	Normal	Multiple Regression
	Not Normal	Kendall Partial Rank (not in WINKS)
	Categorical	Discriminant Analysis (not in WINKS)

Part 4 – BASIC Statistical Procedures

This chapter describes WINKS Basic Procedures. **This chapter assumes you have already gone through the tutorials** in Part 1. The tutorials are the quickest way to get up and running in WINKS. Please go back to Part 1 if you have not yet performed the tutorials.

The analyses described here are accessed through menu items off the WINKS Analyze menu. These include

- Descriptive Statistics
- Graphs and Charts
- t-tests and ANOVA
- Non-Parametric Comparisons
- Regression and Correlation
- Crosstabulations, Frequencies and Chi-Square
- Life Table and Survival Analysis
- Analyze from Summary Data
- Simulations and Demonstrations

For information about additional features included in WINKS Professional, go to Part 5 of this User's Guide.

Detailed Statistics and Histogram
(Analyze/Descriptives/Detail-One Variable)

This option calculates the mean, standard deviation, median, standard error of the mean, minimum, maximum, sum, variance and other descriptive statistics for a single variable (field) from a set of data. For example, suppose you want to calculate statistics for the TIME1 field in the EXAMPLE database.

Step 1: Choose Open Database from the *File* menu. (File icon) Select the EXAMPLE database.

Step 2: From the *Analyze* menu, choose the "Detailed/One Variable" option from the Descriptives sub-menu. (or click the X-bar icon).

Step 3: Choose the TIME1 field to analyze. The results appear in the viewer. Example output is shown below:

```
--------------------------------------------------------------------------
Descriptive Statistics
--------------------------------------------------------------------------
Variable Name is TIME1

N         = 50                Missing or Deleted = 0
Mean      = 21.268               St. Dev (n-1) = 1.71696
Median    = 21.30                St. Dev (n)   = 1.6997
Minimum   = 17.00                    S.E.M.    = 0.24281
Maximum   = 24.20                 Variance     = 2.94793
Sum       = 1063.40              Coef. Var.   = 0.08073
--------------------------------------------------------------------------
Percentiles:                             Tukey Five Number Summary:
0.0%        = 17.00    Minimum            Minimum  = 17.00
0.5%        = 17.00                       25th     = 20.15
2.5%        = 17.3025                     Median   = 21.30
10.0%       = 18.91                       75th     = 22.60
25.0%       = 20.15    Quartile          Maximum  = 24.20
50.0%       = 21.30    Median
75.0%       = 22.60    Quartile
90.0%       = 23.50
97.5%       = 24.1175
99.5%       = 24.20                      Test for normality results:
100.0%      = 24.20    Maximum            D = .093      p >= 0.20

Five number summary consists of the 0, 25, 50, 75 and 100th percentiles.

Confidence Intervals about the mean:
--------------------------------------------------------------------------
80 % C.I. based on a t(49) critical value of 1.3 is (20.95234, 21.58366)
90 % C.I. based on a t(49) critical value of 1.68 is (20.86007, 21.67593)
95 % C.I. based on a t(49) critical value of 2.01 is (20.77994, 21.75606)
98 % C.I. based on a t(49) critical value of 2.41 is (20.68282, 21.85318)
99 % C.I. based on a t(49) critical value of 2.68 is (20.61726, 21.91874)

The normality test suggests that the data are approx. normally distributed.
The test for normality is a modified Kolmogorov-Smirnov test based on
papers by Lilliefors and Dallal & Wilkinson. References in latenews.txt.
```

Click the Graph button to display a histogram of the data. (See the example histogram in Part 1 of this User's Guide.) Definitions of reported statistics:

C. I. - Confidence interval - A range that describes with some confidence where the actual mean of the population from which the data are drawn lies. That is, the true mean (in the example above) is somewhere between 20.79 and 21.23, with 95% confidence.

MAXIMUM - The largest number.

MEAN - The arithmetic average.

MEDIAN - The median is a statistic such that 50% of all numbers in the sample are above the mean and 50% are below the mean.

MINIMUM - The smallest number.

MISSING – Count of data with a missing value code.

N – Sample size.

PERCENTILES - What percent of numbers are lower than that percentile. (50th percentile is the median.)

S.E.M. – (Standard Error of the Mean) measures the precision of the sample mean as an estimate of the population mean.

ST. DEV. (Standard Deviation) - measures the spread of the data around the mean value. It is calculated two ways, using n-1 as a divisor and using n as a divisor. Usually, most people use the n-1 version.

SUM - The total of all the numbers added together

TEST FOR NORMALITY - (Professional edition) - Test that the data are from a normal distribution. The test statistic is D. If the $p <= 0.05$, there is evidence that the data are NOT from a normal distribution.

TUKEY 5 NUMBER SUMMARY - Essentially, the 0th, 25th, 50th, 70th and 100th percentile. See the Hoaglin, et al. reference.

VARIANCE - A measure of the spread of the data -- the square of the st. dev.

Normal Probability Plot

To display a Normal Probability plot first display the Graph from the Detailed Statistics screen, then click on the "Distribution button" (second from left, next to Options button). This button causes the display to cycle through several views of the data, including a probability plot, as shown below.

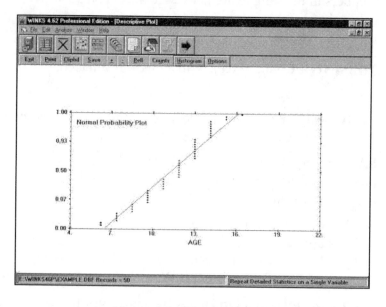

Normal Probability plot.

Detailed Statistics With Cp and Cpk
(Analyze/Descriptives/Detail Cp & Cpk)

This option allows you to perform Cp and Cpk calculations. You will be prompted to enter a lower and upper limit for the data, and the program will use that information to calculate specification limit values.

For example, use the PISTONS.DBF database, and select the Observations variable and specify 73.95 and 74.05 as the lower and upper limits to create the following output (edited):

```
The Process-Capability analysis is used to analyze
process variability relative to product requirements or specifications.

Variable Name is OBSERVED

N        =   125         Missing or Deleted = 0
Mean     = 74.00118        St. Dev (n-1) = 0.01007
S.E.M.   = 0.0009             Variance  = 0.0001
Sum      = 9250.14703      Coef. Var. = 0.00014

Estimated Process Capability Using Mean ± 3S
-------------------------------------------------
74.00118 ± 0.03021, or    (73.97097, 74.03139)

For the normal distribution, the mean ± 3-sigma limits include 99.73%
of the data, and .27% of the data will fall outside these limits.

The following calculations are based on these specification limits:

LSL = 73.95        USL = 74.05

Using S = 0.01007, the estimated PCR (Process Capability Ratio) = 1.6551
95% C.I. on PCR = (1.43687, 1.89928)

The percentage of the specification band used up by the process = 60.4%

Process Fallout = 0.71526 PPM (two-sided defective parts per million)
Process Fallout = 0.35763 PPM (one-sided defective parts per million)

PCR Limits: (PCR(L), PCR(U) = (1.69403, 1.61616)

Fallout based on PCR(L) (one-sided, lower) =0 defective PPM.
Fallout based on PCR(U) (one-sided, upper) =1 defective PPM.

PCRk# (CPk) = 1.61616 (1-sided PCR for the limit nearest process ave.)
```

Summary Statistics on a Number of Variables
(Analyze/Descriptives/Summary-Several Variables)

This option allows you to calculate statistics on several variables (sample size, mean, standard deviation, minimum, maximum, and standard error of the mean). If you have a grouping variable in your database, you may request output of summary statistics by group.

Suppose you want to know the means of all the quantitative variables (AGE, TIME1, TIME2, TIME3, TIME4, STATUS) within each of the three groups (A, B, C) in the EXAMPLE database. Follow these steps:

> **Step 1:** Choose Open Database from the FILE menu. (file icon) Select the EXAMPLE database. Step 2: Choose the Descriptives option from the ANALYZE menu.

> **Step 3**: Choose the "Summary/Several Variables" option from the Descriptives menu.

> **Step 4:** Choose the data fields to analysis: Select the fields AGE, TIME1, TIME2, TIME3, TIME4.

> **Step 5**: Select STATUS as the Group field.

The results viewer will appear displaying summary statistics by STATUS on the group of variables you have selected.

If you do not already have your data in a file, choose New Database from the FILE menu. Then, select the database type you want to create, enter and save the data, and perform the analysis as in the above example.

Percentile Calculation
(Analyze/Descriptives/Percentile Calculation)

This procedure allows you to calculate percentiles for a single variable. You will be asked to select a variable and then enter a list of percentiles to calculate.

Detailed Statistics from Data Entered by Counts
(Analyze/Descriptives/Detailed Statistics from Counts)

If your data is grouped so that you know how many of each number you have (i.e., you have 12 people 13 years old, 5 people

14 yrs old, 6 people 15 yrs old, etc.) you can enter the data by counts.

If your data is already in a database, perform the analysis using the following steps. For example, suppose you want to calculate statistics for the Value field in the COUNTS database.

Step 1: Choose Open Database from the FILE menu. Select the COUNTS database.

Step 2: Choose the Descriptive Statistics option from the Analyze menu.

Step 3: Choose the "Detailed Statistics by Counts" option from the Descriptive Statistics sub-menu.

Step 4: Choose Value as the data field, and Counts as the count field.

If you do not already have your data in a file, choose New Database from the FILE menu. Then, select the database type you want to create (usually "Statistics From Count Data"), enter and save the data, and perform the analysis as in the above example.

The same output as for the Detailed Statistics option is displayed. When the output screen appears, click the Graph button to display a histogram of the data.

Stem and Leaf Display
(Analyze/Descriptives/Stem and Leaf Display)

The Stem and Leaf Display is a graph created from a series of numbers. The Stem part of the display is the leading digit for the data (such as 5 in 54) and the leaf is the trailing digit (such as the 4 in 54). When larger numbers are used, the rightmost digits are often ignored. For example, if the numbers range from 241 to 845, the stem might be the 2 to 8, representing 200 to 800, and the leaf would be 0 to 9, representing the 10's. The 1's place would be ignored. WINKS gives you options for choosing the magnitude of the stem and leaf values. The display below shows a display for DEFLATOR in the LONGLEY database:

```
  4                    8 | 3889
  7                    9 | 689
 (4)                  10 | 0148
  5                   11 | 02456
```

Numbers to the left of the vertical bar "|" are stem values. Digits to the right of the | represents a leaf. The 3 on the first line represents 83. The scale to the left of the stem reports a cumulative count until the stem containing the median is found. The (4) reports which stem contains the median. In this display, we know the median is between 100 and 108. Following the median, the cumulative values count the number of values from the bottom of the display to the median.

For example:

Step 1: Choose Open Database from the FILE menu. (file icon) Select the LONGLEY database.

Step 2: Choose the Descriptives option from the ANALYZE menu.

Step 3: Choose the "Stem and Leaf Display" option from the Descriptives menu.

Step 4: Choose DEFLATOR as the data field to analyze.

Step 5: A dialog box will appear Select 1 for the "letter or digit" and leave the "Split stem value in half" option blank (unchecked).

A Stem and Leaf as show above will be displayed.

Approximate p-value Determination
(Analyze/Descriptives/p-Value Determination)

This option calculates p-values for four test statistics: normal (z), student's t, F, chi-square. Enter the statistic, degrees of freedom and the calculated value of the statistic, and the program will tell you the p-value associated with that statistic. The p-values calculated for the F and chi-square statistics are one-tailed. The p-values calculated for the z and t-statistics are two-tailed. To calculate a p-value, follow these steps:

Step 1: From the Analyze pull-down menu, choose the Descriptive Statistics option.

Step 2: Select Approximate p-value determination.

Step 3: Select what kind of value you want to calculate (t, z, Chi-Square or F) For example, if you want to find the p-value for a t-statistic with 20 degrees of freedom that equals 2.0, select the t-statistic option.

Step 4: For the Statistic Value, enter 2 (press tab) then enter 20 as the degrees of freedom.

Step 5: Select calculate, and the p-value will appear.

Probability Calculator

This option in the p-value calculator allows you to create tables for specific alpha (significance) levels for the z, t, F and Chi-Square distribution. For example, select the probability calculator from the Analyze/Descriptives menu. Select the "Table" button. Choose the Student-t option. Notice the significance level text box. The default values are 0.1 and 0.05; however, you can enter any levels here you want to calculate. Also you can enter what degrees of freedom to use in the table. Click on Start and the table specified by the significance levels and degrees of freedom is calculated and displayed. A table like the following will be displayed:

```
Critical values for Students t-test (one-tail).

DF            .010            .050
---           -----           -----
1             31.83           6.31
2             6.96            2.92
3             4.54            2.35
4             3.75            2.13
5             3.37            2.01
10            2.76            1.81
15            2.60            1.75
```

This table will match closely the critical values table in the back of a statistics text.

Graphs and Charts

The plots described here are stand alone plots. Plots specific to a procedure are described in the section for that feature.

Histogram/Stats
(Analyze/Graphs/Charts – Histogram/Stats)

The Histogram/Stats option in the Graphs menu produces the same output as the Detailed Statistic option in the Descriptive Statistics option. This plot was described in Part 1.

XY Plot (Scatterplot)
(Analyze/Graphs/Charts – X-Y Scatterplot)

An XY plot (scatterplot) displays the relationship between two variables. Such a plot is helpful in determining if two variables are related, and if the relationship is linear (a straight line), curvilinear, or something else.

For example, using the EXAMPLE database, select Time1 and Time2 and select Group for the Group field. The following plot is displayed:

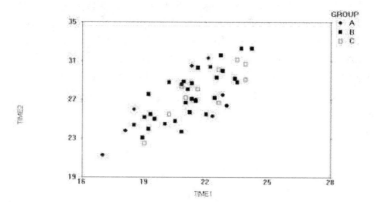

Scatterplot with Groups

From the Options menu you can select items to customize your plot such as an option to display a regression line.

Line Chart
(Analyze/Graphs/Charts – Line Chart)

Select a single variable for a line chart and a graph with connecting lines will be displayed. For example, if you choose the SUN.DBF database (sunspots) the following graph is displayed:

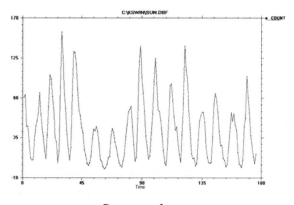

Sunspot data

Correlation Matrix Graphs
(Analyze/Graphs/Charts – Correlation Matrix Graphs)

This option allows you to examine the relationship between several pairs of variables at once. Select the variables you want to compare and a matrix of graphs will be created, as shown below:

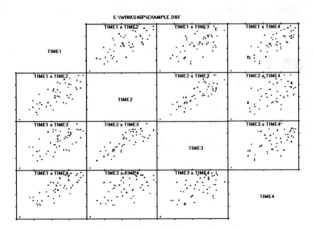

Matrix of Correlation Graphs

By Group Plot
(Analyze/Graphs/Charts – By Group Plots)

By Group Plots allow you to compare distributions of data by groups. The following example compares AGE by GROUP in the EXAMPLE database.

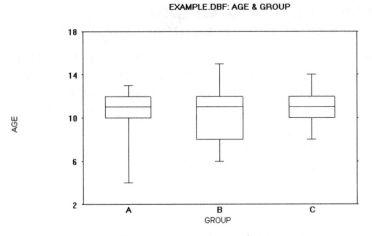

Group comparison plot

First Impression Plots

(Analyze/Graphs/Charts – First Impression Plots)

These plots use a program included with WINKS called First Impression to display charts. See the example First Impression plot in the tutorials in Part 1 of this User's Guide.

First Impression X-Y Scatterplot

The database structure for this plot is the same as for the XY-Scatterplot. However, this plot does not allow a grouping variable. To display this version of the scatterplot, select the XY plot option from the First Impression Gallery. Once a First Impression chart is displayed, press the right mouse button to display menu choices, including printing, capturing, and modifying display options for the chart.

First Impression Bar Chart

A bar chart visually displays the number of counts by group. Thus, a database for this graph consists of a grouping variable (label) and counts. To display a bar chart, follow these steps:

Step 1: Choose Open Database from the FILE menu. Select the BARCHART database.

Step 2: From the Chart Gallery choose Bar Chart.

Step 3: Select one or more data values and a label field. For this example, select VAR1 and VAR2 as data fields, and Label as the label field. Optionally, you can right click and change the chart features such as Plot/3D. The figure below illustrates a bar chart:

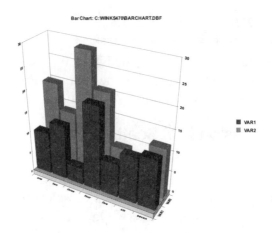

Bar Chart: C:\WINKS470\BARCHART.DBF

First Impression Bar Chart

First Impression Pie Chart

A pie chart is created from a list of counts. A database for a pie chart should contain a Label field and a Value field. The database is similar to a barchart database. To display an example pie chart, follow these steps:

Step 1: Choose Open Database from the FILE menu. Select the BARCHART database.

Step 2: From First Impression Gallery choose Pie Chart from the First Impression Gallery.

Step 3: Select one data value and a label field. For this example, select VAR1 as the data field, and Label as the label field. The figure below shows a 3D Pie Chart.

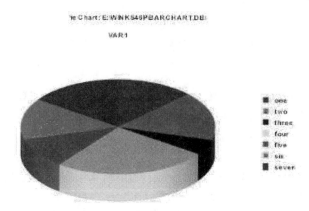

First Impression 3D Pie Chart

First Impression Line/Area Chart

A line, area or time series chart is useful in examining data that are time related, such as profit by month, etc. The X (bottom) axis is assumed to be "time". This plot can also be used as a sequence plot, a plot of the response values in the order in which they are taken. This can be useful to check for trends, perhaps undesirable, in data gathering. A sequence plot showing a trend may call into question the randomness of the sample of observations. On this plot, the points are connected for visual clarity.

To display a time series plot, you should have a database containing one or more VALUE fields and an optional LABEL field. For example, follow these steps to display a chart:

Step 1: Choose Open Database from the FILE menu. Select the EXAMPLE database.

Step 2: From the First Impression Gallery choose Line/Area Chart from the First Impression Gallery.

Step 3: Select one or more data values. For this example, select TIME1 and TIME2 as the data fields.

t-Tests and ANOVA

This series of procedures allows you to compare means across groups.

Independent Group t-test (Analyze/t-Tests and ANOVA – Independent Group t-test/ANOVA)

Independent group analysis is appropriate when observations are taken from groups in which subjects in one group do not appear in another group. That is, the observations within as well as between groups are independent of one another. A t-test is performed when there are two groups, and an ANOVA is performed when there are three or more groups being compared. When performing a t-test or ANOVA on two or more independent groups, you are testing the hypotheses:

Ho: The difference in the means of the groups is zero.
Ha: The difference in the means of the groups is not zero.

For a two-sample t-test, two t-statistics are calculated, one for the case in which the variances of the two samples are equal and the other for use in the case of unequal variances. WINKS performs a test that the variances are equal. If this p-value for this test is small (e.g., less than 0.05), the hypothesis of equal variances is rejected and you use the t-statistic for unequal variances. If the p-value is large, use the t-statistic for equal variances.

Since the observations are all independent of one another, each observation is entered as an individual record in the database. The number of data records must be the same as the total number of observations. Each record includes the response value of one observation and a number or character to indicate to which treatment group it belongs. That is, there will be two fields (variables), one in which to record the response and one in which to indicate the group.

Independent Group t-test Example
(Analyze/t-Tests and ANOVA – Independent Group t-test/ANOVA)

Data are heights of 13 plants grown using two different fertilizers. You want to know if there is a difference in the average heights of plants in the two treatment groups. Data are:

Present Fertilizer	Newer Fertilizer
46.2 cm	51.3 cm
55.6	52.4
53.3	54.6
44.8	52.2
55.4	64.3
56.0	55.0
48.9	

Assign identification to the fertilizer type (i.e., 1 and 2.) Follow these steps to perform this analysis:

Step 1: Since the observations are independent, the database will include thirteen records (one for each plant) and two fields (one for the response and one for the group indicator. Create a database (File/New Database) using the pre-defined structure named "For Independent Group t-test or ANOVA."

This will create a database with the fields GROUP and OBS. Or, use the database on disk named FERTILIZ.DBF and disk to step 3.

Step 2: Enter the data will be in two columns, like this:

Rec. No.	GROUP	OBS (HEIGHT)
1	1	46.2
2	1	55.6
3	1	53.3
4	1	44.8
5	1	55.4
6	1	56.0
7	1	48.9
8	2	51.3
9	2	52.4
10	2	54.6
11	2	52.2
12	2	64.3
13	2	55.0

Step 3: Choose the t-tests and ANOVA option from the Analyze menu, then "Independent group (t-test/ ANOVA).".

Step 4: Select GROUP as the Group variable and OBS (Height) as the data (response) variable. The results are shown below:

```
Group Means and Standard Deviations
------------------------------------
1: mean = 51.4571            s.d. = 4.7476            n =  7
2: mean = 54.9667            s.d. = 4.7944            n =  6

Mean Difference = -3.50952       Pooled s.d. = 2.65538

Test for Equality of Variance
------------------------------
This preliminary test determines which version of the t-test to perform.

Test equality of variance: F = 1.02 with (5, 6) D.F.  p = 0.961 (two-tail)

Note: Since the p-value for equality of variance is greater than 0.05,
use the Equal variance t-test results.

Independent Group t-test Hypotheses
------------------------------------
Ho: There is no difference between means.
Ha: The means are different.

Independent Group t-test on OBS
-------------------------------------------------------------------------
   Equal variance: Calculated t= -1.32 with  11  D.F. p = 0.213 (two-tail)
Unequal variance: Calculated t= -1.32 with 10.7 D.F. p = 0.214 (two-tail)

Confidence Interval
-------------------
A 95% Confidence Interval about the mean difference is: (-9.3541 to 2.335)

Based on a standard error of 2.6554 and a 0.05% t-statistic of 2.201 with 11
d.f.
```

A test for equality of variance (F=1.02) is also performed to see if the variances of the two groups can be considered equal. This is necessary for deciding which t-statistic and p-value to use for the test on means. In this case, p=0.961 indicates that the variances can be consider not different. Since variances are considered not different, use the "Equal variances" t-statistic. Otherwise, you would use the "Unequal variances" result.

The p-value for the appropriate t-test is $p = 0.213$. Since this p-value is greater than 0.05 the decision would be that there is no significant difference between the two groups. (Do not reject the null hypothesis.) Thus, there is not enough evidence to conclude that the means are different and thus there is no significant difference between the groups produced by the treatment or factor.

There is not enough evidence to conclude that the average height of plants grown with the newer fertilizer is significantly different from the average height of plants grown with the present fertilizer.

Step 5: Click the Graph button to display a graphical comparison. This plot shows a box plot comparison of the two groups.

One-Way ANOVA Example
(Analyze/t-Tests and ANOVA – Independent Group t-test/ANOVA)

When more than two independent groups are compared with respect to one variable, one-way or single factor analysis of variance techniques are appropriate. When performing an ANOVA on two or more independent groups, you are testing the hypotheses:

Ho: The difference in the means of the groups is zero.
Ha: The difference in the means of the groups is not zero.

This example uses data for hogs which have been randomly assigned to four groups, with each group being given a different feed. The response is weight gain.

Data for Independent Group ANOVA

Gp 1	Gp 2	Gp 3	Gp 4
60.8	78.7	92.6	86.9
67.0	77.7	84.1	82.2
54.6	76.3	90.5	83.7
61.7	79.8	90.3	

The database to analyze this data is similar to the one used for the t-test example, differing only with respect to the number of groups. In fact, this one-way ANOVA is an extension of the t-test when there are three or more groups. To perform this analysis, use these steps:

Step 1: Create a database (File/New Database) using the pre-defined structure named "For Independent Group t-test or ANOVA." This will create a database with the fields GROUP and OBS. Or, use the database on disk named HOGDATA.DBF and skip to step 3.

The groups will be numbered 1,2,3,4 according to the type of feed used. The response field will be OBS.

Step 2: Enter the data. The data, as entered into the database will look like this:

```
NO GROUP OBS (WEIGHT)
1    1     60.8
2    1     67.0
3    1     54.6
4    1     61.7
5    2     78.7
6    2     77.7
etc...
14   4     83.7
15   4     90.3
```

Step 3: Choose the t-tests and ANOVA option from the Analyze menu, then choose the "Independent group (t-test/ ANOVA)" option.

Step 4: Select GROUP as the Group field and OBS (Weight) as the data field and click Ok. The results appear in the viewer:

```
--------------------------------------------------------------------
Independent Group Analysis                      C:\WINKS\HOGDATA.DBF
--------------------------------------------------------------------
Grouping variable is GROUP
Analysis variable is OBS

Group Means and Standard Deviations

1: mean = 61.025 s.d. = 5.0822 n = 4
2: mean = 78.125 s.d. = 1.4886 n = 4
3: mean = 89.0667 s.d. = 4.4276 n = 3
4: mean = 85.775 s.d. = 3.5976 n = 4

(continues)
```

```
Analysis of Variance Table

Source                   S.S.           DF      MS       F      Appx P
------------------------------------------------------------------------
Total                    1923.41        14
Treatment                1761.24         3     587.08   39.82   <0.001
Error                     162.17        11      14.74

Error term used for comparisons = 14.74 with 11 d.f.


Newman-Keuls Multiple Comp.  Difference P          Q              (.05)
------------------------------------------------------------------------
Mean(3)-Mean(1) =          28.0417    4         13.523        4.256 *
Mean(3)-Mean(2) =          10.9417    3          5.277        3.82  *
Mean(3)-Mean(4) =           3.2917    2          1.587        3.113
Mean(4)-Mean(1) =          24.75      3         12.892        3.82  *
Mean(4)-Mean(2) =           7.65      2          3.985        3.113 *
Mean(2)-Mean(1) =          17.1       2          8.907        3.113 *

Homogeneous Populations, groups ranked

Gp Gp Gp Gp
1   2  4  3
        ------
---
    ---

This is a graphical representation of the Newman-Keuls multiple comparisons
test. At the 0.05 significance level, the means of any two groups
underscored by the same line are not significantly different.
```

Step 5: The results of this test are summarized in the p-value. In this case, the small p-value (p<0.001) means that there is a significant difference between groups. This is taken as evidence of a difference between feeds, a difference not due to chance.

The ANOVA tells you only that there is a difference among the feeds. In order to find out which groups are significantly different from which others, examine the multiple comparison results. The Newman-Keuls multiple comparison test (or whichever you specified at setup) will describe which of the means are significantly different from which others (at the 0.05 significance level).

In the multiple comparison table, comparisons marked with an asterisk "*" are significantly different at the 0.05 significance level (alpha-level). The group numbers are given in increasing order of the value of their group means. That is, Group 1 has the smallest mean, Group 3 the largest. At the 0.05 significance level, the means of any two groups underscored by the same line are not significantly different.

1) The mean for group 1 (feed 1) is statistically significantly less than the means for all other groups.

2) The mean for group 2 (feed 2) is significantly greater than the mean for group 1, and significantly less than the means of groups 4 and 3.

3) The means for groups 4 and 3 are not significantly different from each other, but they are both significantly greater than the means of groups 1 and 2.

You can conclude that feeds 3 and 4 are better than feeds 1 and 2, but there is not enough evidence to say that either feed 3 or 4 is the best overall.

Step 6 Click on the Graph button to view the graphical comparison: This is the same graph as described in the section "By Group Plots." Using the Option button, you can use this graph to show comparison in terms of error bars, box plots, etc.

Independent group test from summary data
(Analyze/t-Tests and ANOVA – Ind. Group from Summary Data)

Consider the fertilizer comparison problem described in the previous independent group t-test example. Suppose you don't have all the data. That is, you aren't given the response of each plant, but you are given the means, standard deviations and sizes of the two groups. You can still perform an analysis.

Summary data for fertilizer study

Group	Mean	St. Dev.	Sample Size
1	51.457	4.748	7
2	54.967	4.794	6

Since you are not given individual data values, you don't have to create a database. Follow these steps to perform an analysis.

Step 1: From the Analyze menu select t-tests and ANOVA. Then select "Ind. Group from summary data."

Step 2: You will be asked for the number of groups. Enter 2. Then you will be asked to enter the mean, standard deviation and sample size for each of the groups. In this case, for group 1 enter 51.457,4.748,7 and press Enter. For group 2 enter 54.967,4.794,6.

Note: Press Tab to move from field to field. Press Enter when you have entered all the data.

Step 3: The interpretation of the results is the same as in the independent group t-test example using this same data.

Note: You can also perform an Independent group one-way ANOVA using summary data. For example, perform an ANOVA using the summary data from the "HOGDATA" example.

Paired and Repeated Measures Analyses

Repeated measures are observations taken on the same or related subjects over time or in differing circumstances. Examples would be weight loss, or reaction to a drug across time. Repeated measures may also be matched subjects.

A t-test is performed when there are two groups (two repeated measures), and an analysis of variance is performed by WINKS if there are more than two groups. The ANOVA determines if there is a difference in the means across groups or repeated measures. A multiple comparison procedure further identifies where the differences lie.

In a database for paired or repeated measures data, each record represents one subject (e.g., person, animal). There must be one field for each repeated measure (each treatment group). For paired data, there are two groups, hence two fields. Thus, in each record, there is a field in which to enter data from each observation (treatment) on that subject.

This repeated measures (paired) analysis requires that all values be available for each subject and any subject with missing values is eliminated from the analysis. That is, a data record must have a value for each field, or it will be eliminated. The hypotheses being tested with a paired t-test or a repeated measures ANOVA is:

> *Ho: There is no difference among means of the groups*
> *(repeated measures).*

> *Ha: There is a difference among means of the groups.*

For comparing matched or paired data (not independent) from two groups, a paired t-test is used. For this test both groups must have the same number of data values, i.e., the two samples must be the same size. A paired t-test is actually a single sample t-test on the mean of the differences between the two data values in each pair.

Paired t-test Example (Repeated Measures)
(Analyze/t-Tests and ANOVA – Paired Rep. Measures (t-test/ANOVA))

In a before/after study, the difference "after minus before" (or "before minus after") is taken for each pair and the average of these differences calculated. If this average of differences is found by a single sample t-test to be significantly different from zero, the conclusion is that there is a change. The data in this example are before and after weights for eight persons on a diet. Notice that in this case, both data values are taken from the SAME entity (person). Follow these steps to perform this analysis:

Step 1: Create a database with two fields (BEFORE and AFTER) and eight records, one for each person. Use the pre-defined database structure named "Paired t-test or McNemar's Test." This will create a database with the fields REP1 and REP2. The REP1 will be used for Before and REP2 will be used for After.

Of course, you can choose to create a custom database and enter a structure containing the fields named BEFORE and AFTER. (Or, open the database named DIET, and skip to step 3.)

The data for the paired t-test is:

Subj	Before	After
1	162	168
2	170	136

```
        3        184        147
        4        164        159
        5        172        143
        6        176        161
        7        159        143
        8        170        145
```

Step 2: Enter the data for the eight records. The database should look similar to the listing of the data above.

Step 3: Choose the t-tests and ANOVA option from the Analyze menu. Choose the "Paired/Rep. Measure (t-test/ANOVA) option.

Step 4: Select REP1 (BEFORE) as the first field and REP2 as the second field and click Ok.

Step 5: The results are shown in the viewer. A portion of the output is shown below:

```
Means and standard deviations for 2 repeated measures:

1)REP1: mean = 169.625 s.d. = 8.07001
2)REP2: mean = 150.25 s.d. = 11.04213

Mean Difference = 19.375 s.d.(difference) = 14.78356

95% C.I. about Mean Difference is (7.01367, 31.73633)

Paired t-test
--------------

Ho: The mean difference between pairs is 0.
Ha: The mean difference between pairs is not 0.

Calculated t = 3.70687 with 7 D.F. p = 0.0076 (two-sided)

Since p <= 0.05, at the 0.05 significance level you have evidence to reject
the null hypothesis and conclude that the mean difference between pairs is
not 0.

For a one-sided test, you must adjust the p-value according to the direction
of your alternative hypothesis.
```

The means and standard deviations for each group are reported, but more importantly, the mean difference between BEFORE and AFTER measurements is given. The statistical procedures are performed on this average difference. These results are interpreted like those of a single sample t-test with null hypothesis: mean=0, and alternative hypothesis: mean <> 0. The calculated t-statistic is 3.70687. The test is performed with 7 degrees of freedom, and the p-value associated with the test is 0.0076. A small p-value such as this is indicates rejection of the null hypothesis and leads to the conclusion that the average difference in BEFORE and AFTER weights is not zero, i.e., there is evidence of a significant (at the 0.05 level) change of weight in these eight subjects on average.

Step 6: Click Graph to display a graphical comparison.

One-way repeated measures ANOVA Example
(Analyze/t-Tests and ANOVA – Paired Rep. Measures (t-test/ANOVA))

For more than a pair of repeated measures, a one-way repeated measures analysis of variance is appropriate. The data in this example are repeated measures of reaction times of five persons after being treated with four drugs in randomized order.

One-way repeated measures ANOVA data

Drug1	Drug2	Drug3	Drug4
31	29	17	35
15	17	11	23
25	21	19	31
35	35	21	45
27	27	15	31

To perform this analysis, follow these steps:

Step 1: Create a database with four fields (DRUG1, DRUG2, DRUG3 and DRUG4) Use the pre-defined database structure named "Paired t-test or McNemar's Test." Or, open the database named DRUG.DBF and disk to step 3.

Step 2: For the first record, enter the data for the first person 31,29,17,35. The second record will contain 15,17,11,23 and so forth. After you finish entering your data, the database should look similar to the data listing above.

Step 3: Choose the t-tests and ANOVA option from the Analyze menu. Select the "Paired/Rep. Measure (t-test/ANOVA) option.

Step 4: Select DRUG1, DRUG2, DRUG3 and DRUG4 as the fields to use and click Ok.

Step 5: The results are displayed in the viewer: Partial output is shown below:

```
Means and standard deviations for 4 repeated measures:

1)DRUG1: mean = 26.6 s.d. = 7.53658
2)DRUG2: mean = 25.8 s.d. = 7.01427
3)DRUG3: mean = 16.6 s.d. = 3.84708
4)DRUG4: mean = 33.0 s.d. = 8.0

Repeated Measures Analysis of Variance

Source              -----S.S.-----   --DF--    MS      F     Appx p
-----------------------------------------------------------------------
Between Subject          648.00        4
Within Subject           775.00       15
  Rep. Factor    683.80                3     227.93  29.99    <.001
  Error           91.2                12       7.60
-----------------------------------------------------------------------
Total                   1423.00       19

Error term used for comparisons = 7.6 with 12 d.f.

                                                        Critical q
Newman-Keuls Multiple Comp. Difference  P      Q          (.05)
-----------------------------------------------------------------------
Mean( 4)-Mean( 3) =          16.4      4    13.302       4.199 *
Mean( 4)-Mean( 2) =           7.2      3     5.84        3.773 *
Mean( 4)-Mean( 1) =           6.4      2     5.191       3.082 *
Mean( 1)-Mean( 3) =          10.0      3     8.111       3.773 *
Mean( 1)-Mean( 2) =           0.8      2      .649       3.082
Mean( 2)-Mean( 3) =           9.2      2     7.462       3.082 *

Homogeneous Populations, repeated measures ranked
Gp 1 refers to DRUG1
Gp 2 refers to DRUG2
Gp 3 refers to DRUG3
Gp 4 refers to DRUG4

        Gp      Gp      Gp      Gp
        3       2       1       4
                ----------
        -----
                        -----
```

In this case, the small p-value (The p<0.001 on the "Repeated Factor" line in the ANOVA table.) means that there is a statistically significant difference in the mean response times for the four drugs.

In the multiple comparison table, comparisons marked with an asterisk "*" are significantly different at the 0.05 significance level (alpha-level).

The graphical comparison (groups underlined) of the differences tells you that groups 2 and 1 (DRUGS 2 and 1) are not significantly different at the 0.05 significance level. The Group 3 (DRUG 3) mean is significantly smaller than all other groups' and the Group 4 (DRUG 4) mean is significantly larger than all others.

Step 6: Click on the Graph button to view a comparison graph.

Single Sample t-test Analysis
(Analyze/t-Tests and ANOVA – Single Sample t-test)

The single sample analysis allows you to choose a single variable, and test a hypothesis that the mean differs from an hypothesized mean. You must enter the hypothesized population mean. The hypotheses you are testing in this case are:

Ho: The mean equals the hypothesized value.
Ha: The mean does not equal the hypothesized value.

Follow these steps to perform the analysis:

Step 1: Choose the Open Database option from the FILE pull-down menu, then choose the EXAMPLE database from the displayed list. Or, create a database with at least one numeric field.

Step 2: From Analyze, choose the t-test and ANOVA option. Then, choose the "Single sample t-test" option.

Step 3: Choose the field you wish to use. In this case, choose the field TIME1.

Step 4: WINKS will display the mean, standard deviation and sample size of the field TIME1, and will ask you for the hypothesized mean. In this case, you want to know if the mean is 20. Enter 20 as the hypothesized mean. Choose Calculate and End.

Step 5: The results of the analysis will be displayed.

```
-------------------------------------------------------------------
Single Sample t-test C:\WINKS\EXAMPLE.DBF
-------------------------------------------------------------------

Variable Name is TIME1

N = 50 Missing or Deleted = 0
Mean = 21.268 St. Dev (n-1) = 1.71695

Null Hypothesis: mean(POPULATION) = 20

Calculated t = 5.22 with 49 D.F. p = < 0.001 (2 - sided test)
```

```
95% C.I. about Mean is (20.77995, 21.75605)
```

A small p-value, such as the one in this case p<0.001, supports rejection of the null hypothesis. That is, you can conclude that the mean of TIME1 in the population is statistically significantly different from 20 based on this test procedure.

Note: Even if you don't have all the data values, you can test whether the mean of a population is equal to some hypothesized value if you know the mean, standard deviation and sample size of the data values in a sample. In this case, choose the Single sample t-test - from summary data option.

Single Sample t-test/Summary Data
(Analyze/t-Tests and ANOVA – Single Sample t-test Summary Data)

If you only have the mean, standard deviation and sample size for a single sample t-test, select this option and you will be prompted to enter the appropriate information. Results will be the same as in the previous example.

Dunnett's Test Single Sample t-test Analysis
(Analyze/t-Tests and ANOVA – Dunnett's Test ANOVA)

Dunnett's test is a multiple comparison procedure following a one-way ANOVA either from database data or from summary data that compares a control mean with the other means in the analysis. Data entry is the same as for a One-Way Independent Group ANOVA. You will be asked to specify which field in represents the control group, and multiple comparisons will be performed accordingly.

Dunnett's Test Single Sample t-test Analysis from Summary Data
(Analyze/t-Tests and ANOVA – Dunnett's Test ANOVA Summary Data)

This option allows you to enter the data as means, standard deviations and sample sizes. Results will be the same as in the previous example.

Non-Parametric Procedures

Non-Parametric procedures are useful as a substitute for parametric tests when you cannot make the assumption that the data follow a normal distribution. It is also useful if you do not have exact data values for the observations but you do have order statistics, that is, you don't know the exact response values but you know which is largest, next largest, and so forth, to smallest.

Mann-Whitney Example
(Analyze/Non-Parametric Comparisons/Ind. Gp. Non-Parametric)

The Mann-Whitney test is **similar to the independent group t-test** except there is no assumption about normality or equality of variance.

The hypotheses being tested are:

   Ho: There is no difference in the medians of the groups.
   Ha: There is a difference in the medians of the groups.

For this example we'll use the same data as in the previous t-test example. Follow these steps to do this example:

Step 1: Open the database named FERTILIZ (or create it as described in the t-test example) and choose the Non-Parametric Comparisons option from the Analyze menu.

Step 2: From the Non-Parametrics Comparisons menu select "Ind. Grp - Mann-Whitney, Kruskal-Wallis."

Note: Critical Values of the Mann-Whitney U Distribution at 0.05 (two tailed) are found in the printed manual.

Step 3: Select GROUP as the group field and OBS (Height) as the data field and choose Ok. The results will be displayed in the viewer. The results are:

```
Group variable = GROUP    Observation variable = OBS

Mann-Whitney U' = 24.    U = 18.

Rank sum group 1 = 46.              N = 7      Mean Rank = 6.57

Rank sum group 2 = 45.              N = 6      Mean Rank = 7.5

Significance estimated using the z statistic.

Z = .357      p = 0.721
```

In this case, U'=24.00, U = 18, z=0.357 and p=0.721. The p-value of 0.721 is large so the null hypothesis of no difference in medians between groups is not rejected. There is not sufficient evidence to say that there is a difference between the median heights of plants in the two groups.

Kruskal-Wallis Example
(Analyze/Non-Parametric Comparisons/Ind. Gp. Non-Parametric)

If **more than two independent groups** are being compared using non-parametric methods, WINKS uses the Kruskal-Wallis test. The data are ranked and summed as in the Mann-Whitney procedure. The hypotheses being tested are:

Ho: There is no difference in the medians of the groups.
Ha: There is a difference in the medians of the groups.

The Kruskal-Wallis procedure tests for a difference among several treatment groups, but does not identify where the difference lies. WINKS performs a multiple comparison test to identify which groups are different from which others. Note: If any n's are less than 5 and k is 3, Kruskal-Wallis tables of critical values (available in most statistical analysis non-parametric texts) should be used. When k is 2, the test reduces to the Mann-Whitney test.

The data used in this example are weights of four groups of seven randomly assigned animals, each group given a different feed treatment. You want to test whether there is a difference among the different treatments. The data for Kruskal-Wallis Procedure is:

Group1	Group2	Group3	Group4
50.8	68.7	82.6	76.9
57.0	67.7	74.1	72.2
44.6	66.3	80.5	73.7

```
51.7    69.8    80.3    74.2
48.2    66.9    81.5    70.6
51.3    65.2    78.6    75.3
49.0    62.0    76.1    69.8
```

Step 1: Since the groups are independent, the database will include two fields (TREATMENT and WEIGHT) and 28 records (one for each animal). Create a database by using pre-defined structure "Independent group t-test and ANOVA." This will create a database with the fields GROUP and OBS. The GROUP variable will be used for the Treatment type (1, 2, 3, or 4) and the OBS will be used for the Weight. Of course, you can choose to create a custom database and enter a structure containing the fields named TREATMENT and WEIGHT. This database is similar to the one used in the independent group ANOVA previously described. Or, open the KRUSKAL.DBF database and skip to step 3.

Step 2: The data you will enter in the first record is 1 (for Group 1) and 50.8. Enter the data for the 28 records. For Example:

```
GROUP  OBS  (TREATMENT)
1         50.8
1         57.0
1         44.6
1         51.7
1         48.2
etc...
4         75.3
4         69.8
```

Step 3: From the Non-Parametrics Comparisons module menu, select "Ind. Gps - Mann-Whitney, Kruskal-Wallis"

Step 4: Choose GROUP (TREATMENT) as the Group variable and choose OBS (WEIGHT) as the data (response) variable.

Step 5: WINKS will display the Kruskal-Wallis H-statistic, the rank sums, sample sizes and mean ranks of the groups, a chi-square statistic and an approximate p-value.

A low p-value (less than the significance level (e.g., less than 0.05) indicates rejection of the null hypothesis. In this case, p<0.001 so the null hypotheses is rejected. That is, there is enough evidence to say that the groups have different medians, i.e., the groups are not identical with respect to location. If you reject the null hypothesis and conclude that the groups have different medians, you may also wish to know which groups differ. Results are:

```
Group variable = GROUP Observation variable = OBS

Kruskal-Wallis H = 24.48

P-value for H estimated by Chi-Square with 3 degrees
of freedom.

Chi-Square = 24.5 with 3 D.F. p < 0.001

Rank sum group 1 = 28. N = 7 Mean Rank = 4.
Rank sum group 2 = 77.5 N = 7 Mean Rank = 11.07
Rank sum group 3 = 171. N = 7 Mean Rank = 24.43
Rank sum group 4 = 129.5 N = 7 Mean Rank = 18.5
```

Tukey Multiple Comp.	Difference	Critical q Q (.05)
Rank(3)-Rank(1) = 20.4286 (SE used = 4.3964)	4.647	2.639 *
Rank(3)-Rank(2) = 13.3571 (SE used = 4.3964)	3.038	2.639 *
Rank(3)-Rank(4) = 5.9286 (SE used = 4.3964)	1.349	2.639
Rank(4)-Rank(1) = 14.5 (SE used = 4.3964)	3.298	2.639 *
Rank(4)-Rank(2) = 7.4286 (SE used = 4.3964)	1.69	2.639
Rank(2)-Rank(1) = 7.0714 (SE used = 4.3964)	1.608	2.639

The multiple comparison procedure based on ranks performs a test to find specific differences. Comparisons marked with an asterisk "*" are significantly different at the 0.05 significance level. A graphical description of the multiple comparisons is given by:

```
      Gp   Gp   Gp   Gp
       1    2    4    3
                 ------

           ------

  ------
```

Groups are listed in increasing average ranks. The conclusion is
that the median of each group is significantly different from every
other median (at the 0.05 significance level).

Wilcoxon Signed Rank Test – Paired Data
(Analyze/Non-Parametric Comparisons/Paired Rep. Measures Non-
Parametric)

Using the previous DIET data from the previous paired t-test, the
Wilcoxon test yields the following results:

```
Sum of the positive ranks = 34.
Sum of the negative ranks = 2.
Number of samples =  8

Using Wilcoxon table lookup, p  = .012 (one-tailed)
Using Wilcoxon table lookup, p  = .024 (two-tailed)
```

The two-tailed test (or one-tailed) indicates that there is a
significant difference in the two groups indicating that there was a
significant weight loss. Sign test results are also reported.

Friedman's Test - Repeated Measures
(Analyze/Non-Parametric Comparisons/Paired Rep. Measures Non-
Parametric)

One method of performing a non-parametric one-way analysis of
variance (ANOVA) with repeated measures (randomized complete
block experimental design) is with the Friedman test.

The hypotheses for the Friedman test are:

Ho:There is no difference in mean ranks between repeated
measures.
Ha:There is a difference in mean ranks between repeated
measures.

The Friedman analysis differs from a standard (parametric repeated
measures) ANOVA in that the analysis is performed on the ranks
of the data rather than on the actual data. For example, the
following data are the same data used in a previous example for a
standard repeated measures ANOVA:

Drug1	Drug2	Drug3	Drug4
31	29	17	35
15	17	11	23
25	21	19	31
35	35	21	45
27	27	15	31

The data presented here are repeated measures of reaction times of 5 persons after being given 4 drugs in randomized order. (See Winer, page 301 for more details.)

The database for this analysis will include four fields (the repeated measures) and will have five records, with each record representing a subject. Follow these steps to perform this analysis:

Step 1: Open the database named DRUG.

Step 2: From the Analyze menu, choose Non-Parametric Comparisons. The Non-Parametric module menu appears. Choose the "Repeated Measures Analysis" option.

Step 3: Choose the fields DRUG1, DRUG2, DRUG2 and DRUG4 (in that order) to use for the analysis and click Ok.

Step 4: A Chi-Square value of 14.13 with p<0.01 is reported. The small p-value means that there is a statistically significant difference in the mean ranks of times for the four drugs. In the multiple comparison table, comparisons marked with an asterisk "*" are significantly different at the 0.05 significance level.

```
Number of repeated measures= 4 Number of subjects = 5

1 )DRUG1 Rank sum = 13.0 Mean rank = 2.6
2 )DRUG2 Rank sum = 12.0 Mean rank = 2.4
3 )DRUG3 Rank sum = 5.0 Mean rank = 1.0
4 )DRUG4 Rank sum = 20.0 Mean rank = 4.0

Ho:There is no difference in mean ranks for repeated measures.
Ha:A difference exists in the mean ranks for repeated measures.

Friedman's Chi-Square = 14.13 with d.f. = 3 p = 0.003

Kendall's coefficient of concordance = 0.942

When the p-value is low, there is evidence to reject Ho,
and conclude that there is a difference between mean ranks.

Error term used for comparisons = 2.89
                                          Critical q
```

```
Tukey Multiple Comp.                Difference      Q (.05)
----------------------------------------------------------------------
Rank( 4)-Rank( 3)  = 15.0           5.196           3.63 *
Rank( 4)-Rank( 2)  = 8.0            2.771           3.63
Rank( 4)-Rank( 1)  = 7.0                            (Do not test)
Rank( 1)-Rank( 3)  = 8.0            2.771           3.63
Rank( 1)-Rank( 2)  = 1.0                            (Do not test)
Rank( 2)-Rank( 3)  = 7.0                            (Do not test)

Homogeneous Populations, repeated measures ranked

Gp 1 refers to DRUG1
Gp 2 refers to DRUG2
Gp 3 refers to DRUG3
Gp 4 refers to DRUG4

Gp Gp Gp Gp
 3  2  1  4
    ---------
---------
```

This above graph is interpreted in the following way: Any two
groups underlined by the same line are considered not different at
the 0.05 level of significance. Therefore, the result of this analysis
is that the mean rank for DRUG 3 is less than the mean rank for
DRUG 4. There are no other statistically significant pair wise
differences among the four groups.

Cochran's Q -Non-Parametric Dichotomous Data Analysis

(Analyze/Non-Parametric Comparisons/Dichotomous Data – Cochran's Q)

Cochran's Q procedure is a non-parametric procedure appropriate for use with dichotomous data when the experiment involves repeated measures on blocks. Often the blocks are subjects (people or animals). The response of the subjects to the treatments is dichotomous if it is taken as one of only two possible outcomes, often labeled "success" and "failure", rather than as a measurement.

Cochrans' Q is used to test three or more treatments, or groups, and is in fact an extension of McNemar's test for two groups (see Crosstabulation procedures). (Cochran's Q test is placed under the Non-Parametric procedures rather than in the Crosstabulation procedure because its data entry requirements fit better in this section.) Cochran's Q can also be seen as similar to Friedman's test when data are dichotomous. The hypotheses being tested are:

Ho: The proportion of successes is the same for all treatments.
Ha: The proportion of successes is not the same for all treatments.

The data can be organized in a table of r rows and c columns, where the rows are the subjects and the columns are the treatments. Each row by column entry of 0 (failure) or 1 (success) represents the response of that row's subject to that column's treatment. For example, in the data used below, you can see that Drug 4 failed on Person 2.

Consider an experiment where the response is not a measurement of a reaction, but instead is simply a "success" or "failure". That is, an experiment is conducted in which six persons are each given five test drugs (say headache remedies) in random order, and a response of "success" or "failure" is recorded in each case. The data are as follows:

Dichotomous data for Cochran's Q (headache study)

Drug1	Drug2	Drug3	Drug4	Drug5
1	1	0	1	0
0	0	0	0	1
0	0	0	1	0
1	1	0	1	1
0	0	0	1	1
1	0	0	1	1

where 1 = success and 0 = failure. Follow these steps to perform an analysis:

Step 1: Open this data is in a file named COCHRAN on disk

Step 2: Select the fields DRUG1, DRUG2, ... DRUG5 as the fields to use in the analysis. The results of the calculations appears in the viewer and click Ok.

Step 3: WINKS summarizes the test information and reports

```
Chi-Square = 9.87  with d.f. = 4  p = 0.044
```

In this case, the p-value is small. Therefore, the null hypothesis is rejected in favor of the alternative hypothesis that there is difference among the headache remedies.

Regression and Correlation

Simple linear regression is used for predicting a value of a dependent variable using an independent variable. Multiple regression is used for predicting the value of a dependent variable using one or more independent variables. Correlation is used to measure the strength of association between two variables. For example, you may be interested in relating advertising to orders received. The question you are asking is, "Is there a relationship between the amount of money spent on advertising and the amount of orders received?" It is also possible to compare more than two variables at a time using multiple regression. For example, you may be interested in how the combination of radio advertising costs, direct mail costs and commissions relate to the number of orders received.

Both regression and correlation measure the linear relationship between the variables. In the case of the Spearman's correlation the relationship measured is an association between the ranks of the data. When the data are plotted (scatterplot), highly associated variables should fall "scattered" about a straight line. You should check this assumption using the scatterplot or residual plot options available in the WINKS regression and correlation procedures.

Regression procedures also assume that for a fixed X value (a fixed value of the independent variable), the population of Y values (values of the dependent variable) is normally distributed and that all these normal distributions have equal variances. You can use the residual plot options on the regression procedures to check this assumption. If the residuals plotted against an independent variable show a pattern other than a band of points randomly scattered about zero, these assumptions may be violated.

Simple Linear Regression Example
(Analyze/Regression and Correlation/Simple Linear Regression)

Data for this example of simple linear regression are Homicide Rate and Handgun Licenses Issued per 100,000 population for the years 1961 to 1973 in Detroit (Fisher, 1976, reprinted from Gunst and Mason, 1980). (HANDGUNS.DBF)

Year	Homicide Rate	Handguns Registered
1961	8.60	178.15
1962	8.90	156.41
1963	8.52	198.02
1964	8.89	222.10
1965	13.07	301.92
1966	14.57	391.22
1967	21.36	665.56
1968	28.03	1131.21
1969	31.49	837.60
1970	37.39	794.90
1971	46.26	817.74
1972	47.24	583.17
1973	52.33	709.59

To compare the homicide rate with handguns registered, you need a database with only these two sets of numbers (you can exclude

year.) The data for this example is stored on your disk as
HANDGUNS.DBF with the variables HOMICIDES and
HANDGUNS. To perform a simple linear regression using this
data, follow these steps:

Step 1: Open the database named HANDGUNS.DBF. Or, you can
create a database using the pre-defined database description
"Simple Linear Regression."

Step 2: From the Analyze menu, select Regression & Correlation
and "Simple Linear Regression."

Step 3: Select HOM_RATE as the DEPENDENT (Y) variable
first, then select HAND_REG as the INDEPENDENT (X)
variable.

Step 4: WINKS displays preliminary results, and asks if you want
to Predict or Continue. The Predict option is used if you want to
use the regression equation to calculate new values of Y by
entering values of X. For this example, select Continue. Regression
results will be displayed in the viewer.

```
Dependent variable is HOM_RATE, 1 independent variables, 13 cases.
-----------------------------------------------------------------
Variable     Coefficient  St. Error   t-value     p(2 tail)
-----------------------------------------------------------------
Intercept    4.9105126    6.6274622   .7409341    0.474
HAND_REG     .0376114     .0107324    3.5044807   0.005
-----------------------------------------------------------------
R-Square = 0.5275       Adjusted R-Square = 0.4846

Analysis of Variance to Test Regression Relation

Source              Sum of Sqs  df   Mean Sq     F           p-value
-------------------------------------------------------------------
Regression          1699.557    1    1699.557    12.281385   0.005
Error               1522.2328   11   138.3848
-------------------------------------------------------------------
Total               3221.7897   12

A low p-value suggests that the dependent variable HOM_RATE may be linearly
related to independent variable(s).

-------------------------------------------------------------------
MEAN X = 537.507    S.D. X = 316.415     CORR XSS = 1201423.0
MEAN Y = 25.127     S.D. Y = 16.385      CORR YSS = 3221.788
REGRESSION MS= 1699.557             RESIDUAL MS= 138.385
-------------------------------------------------------------------

Pearson's r (Correlation Coefficient)= 0.7263

The linear regression equation is:
```

```
HOM_RATE = 4.910512 + 3.761144E-02 * HAND_REG

Test of hypothesis to determine significance of relationship:
        H(null): Slope = 0 or H(null): r = 0 (two-tailed test)
        t = 3.5 with 11 degrees of freedom p = 0.005

Note: A low p-value implies that the slope does not = 0.
```

The table at the top of the output tells you the intercept value and the coefficient values for each of the independent variables. These can be used to create a prediction equation as explained below. Pearson's correlation coefficient (r) is reported (0.7263). Pearson's r ranges from -1 to 1; the further r is from 0, the stronger the correlation. In this case, r=0.7263, not necessarily a strong correlation, but not weak either although the test of hypothesis determines that r is significantly different from zero (p=0.005). How substantial a correlation of this strength is depends on the situation and the judgment of the researcher. R^2 ranges from 0 to 1; the closer to one (1) R^2 is, the better fit the regression line is to the data. The linear regression equation given is a mathematical representation of a straight line that passes through a plot of the data, and can be used to predict the dependent variable (HOMICIDES) given a value for the independent variable (HANDGUNS). In this case the linear regression equation is:

```
HOMICIDES = 4.910512 + 3.761144E-02 * HANDGUNS
```

(The E notation is scientific. Thus, 3.7E-02 means 0.037.) If you want to predict the homicide rate for 300 handguns registered, you would use the equation:

```
HOMICIDES = 4.910512 + 3.761144E-02 * 300
```

A t-test is performed to test the statistical significance of the linear relationship between the two variables. A low p-value means that the two variables are significantly related. In this case p=0.005, quite small, so the null hypothesis (Slope = 0) is rejected and you conclude that the regression line has a slope significantly different from zero. That is, there is a significant linear relationship between homicides and handguns for the years 1961 to 1973 in Detroit, and within the range 178 to 1131 handguns. Reference a text on regression for warnings about how to use (or not to use) this kind of information for prediction purposes.

Step 5: Click Graph to view regression plots. You may choose to view a scatterplot of the original data with the fitted regression line (Regression Plot), or a plot of the residual values by choosing from the combo box displayed, then Ok.

Plots are helpful in visually examining the relationship between the variables. It is important to verify that the relationship is indeed a straight line. If it is not, a non-linear pattern should emerge from the scatterplot of the data, or a pattern other than a random horizontal band centered at zero from the plot of the residuals. See Neter and Wassermen's book for a good description of residual plots (see references). If the relationship is not linear, they could possibly be transformed to make the data linear.

Prediction Intervals in Simple Regression

After performing a simple linear regression, you can select the "Predict" button on the text viewer to calculate predicted "Y" values using the regression equation. Beginning with version 4.5, the predicted Y values are accompanied by a 95% prediction interval on those values.

Regression Through the Origin

A standard simple linear regression procedure calculates coefficients for an intercept and slope term in the linear equation. However, sometimes your knowledge about the true nature of the relationship between your two variables tells you that the intercept coefficient should be 0 (zero). That is, the line should pass through the origin. When this is the case, you can force WINKS to estimate the best slope of the line (least squares) that will fit a straight line through the origin and fit the scatter of points. The following example illustrates how to perform a linear regression with a forced zero intercept.

Step 1: Open the database named ORIGIN.DBF. This database contains two variables, VAR1 and VAR2.

Step 2: From the Analyze menu, select "Regression and Correlation" then "Simple Linear Regression."

Step 3: You will be prompted to select the dependent and independent variables for the regression. In this case, select VAR1 as the Independent variable and VAR2 as the Dependent variable.

Step 4: Important: Before clicking the Okay button, **click on the radio button labeled "Regression through origin option."**

The output that will be displayed is:

```
-------------------------------------------------------------------
Dependent variable is VAR2, 1 independent variables, 12 cases.
-------------------------------------------------------------------
Variable Coefficient St. Error t-value p(2 tail)
-------------------------------------------------------------------
VAR1 4.6852741 .034205 136.9762 <.001

A 95% Confidence (using t(.975, 11) = 2.201 interval for
VAR1 is:

( 4.609988, 4.76056)
The estimated regression function is:

VAR2-hat = 4.6852741106916 * VAR1
```

Notice that the normal intercept term is missing. The 4.68 value is the slope of the line though the scatter of points. A t-test yields a t-value of 136.97 and a p-value < 0.001. This tells you that the slope is significantly different from zero. The 95% confidence interval tells you that the slope is most likely between about 4.609988 and 4.76056.

The use and interpretation of the resulting equation is similar to that for a Simple Linear Regression.

Multiple Regression Example
(Analyze/Regression and Correlation/Multiple Linear Regression)

Multiple regression is an extension of simple linear regression into several dimensions (several independent variables). In the multiple regression procedure, you must enter a list of the independent variables and a single dependent variable on which you wish to perform the regression analysis. In WINKS you may use up to 20 independent variables in this option. Multiple regression can be complicated. Refer to a good text on the subject before making any conclusions about your results.

WINKS calculates and displays several results, including the coefficients and intercept of the regression "line". A significance test is performed to determine the significance of the contribution of the different variables or factors to the model (mathematical representation). Also displayed is R-square (R^2), as well as adjusted R-square. R-square varies from 0.0 to 1.0, with 0.0 meaning no relationship (model is not good) and 1.0 meaning the regression equation perfectly describes the sample data.

An analysis of variance is performed to determine the overall significance of the model. If the ANOVA reveals a significant relationship, (that is, if the p-value is small) the model may be a good representation of the sample data. A plot of residuals from the fit is available. You may plot the fit against any of the variables. Look for patterns in the residuals. Patterns other than a horizontal band about zero suggest that the assumptions necessary for regression analysis may be violated.

Longley (1967) introduced a data set which has often been used in comparing multiple linear regression procedures in the literature. The variables refer to economic factors. This example uses the LONGLEY database on the WINKS disk. Follow these steps to perform a multiple linear regression:

Step 1: Open the database named LONGLEY.

Step 2: From the Analyze menu, select "Regression and Correlation," then "Multiple Regression Analysis"

Step 3: The LONGLEY database consists of 7 fields. Select TOTAL as the DEPENDENT variable and DEFLATOR, GNP, UNEMP, ARMED, POP, TIME as the INDEPENDENT variables, then click on Ok.

Step 4: WINKS will display preliminary results, and ask if you want to Predict or Continue. The Predict option is used if you want to use the regression equation to calculate new values of Y by entering values of X. For this example, select Continue.

Regression results will be displayed in the viewer.

```
Dependent variable is TOTAL, 6 independent variables,
16 cases.
------------------------------------------------------------------
Variable   Coefficient    St. Error    t-value    p(2 tail)
------------------------------------------------------------------
Intercept  -3482258.6349  889652.92    -3.914177  0.004
DEFLATOR   15.061872      84.841736    .177529    0.863
GNP        -.0358192      .0334621     -1.070439  0.312
UNEMP      -2.02023       .4879787     -4.139996  0.003
ARMED      -1.033227      .2140895     -4.826145  <.001
POP        -.0511041      .2258783     -.2262461  0.826
TIME       1829.1515      455.08592    4.0193541  0.003
------------------------------------------------------------------
R-Square = 0.9955 Adjusted R-Square = 0.9925

Analysis of Variance to Test Regression Relation

Source Sum of Sqs df Mean Sq F p-value
------------------------------------------------------------------
Regression 184173843.173 6 30695640.5288 330.85802 <.001
Error 834982.83 9 92775.87
------------------------------------------------------------------
Total 185008826. 15
```

```
A low p-value suggests that the dependent variable
TOTAL may be linearly related to independent
variable(s).
```

The table at the top of the output tells you the intercept value and the coefficient values for each of the independent variables. These can be used to create an equation for prediction of the dependent variable. In this case, the equation is:

```
TOTAL = -3481930.1065 + DEFLATOR*(15.0161517122) +
GNP*(-0.03579443400) + UNEMP*(-2.0199053296) +
ARMED*(-1.0332049046) + POP*(-0.05130725587) +
TIME*(1828.99249535)
```

Note: *Although the results are reported to 8 to 9 decimal places, it is usually not appropriate or necessary to use this many decimal places.*

The t-value associated with each coefficient tests its significance in the equation. You can use the p-value associated with each coefficient to make a decision about the validity of having that variable in the equation. A low p-value suggests that the dependent variable, TOTAL, is related to the independent variable whose p-value you are examining. In this case, you might question the validity of having DEFLATOR (p=0.8636), GNP (p=0.3132) and POP (p=0.8257) in the equation.

In choosing the variables to have in such an equation, you also need to consider such questions as multicollinearity, heteroscedasticity and parsimony. There are also other ways to approach the selection of variables for a multiple regression equation. Refer to a good text on regression. If you wish to delete some variables from the equation, you can do so by redoing the analysis and leaving some of the variables out of the equation.

WINKS also reports R-Square, which gives you a measure of how well the regression "line" fits the data, and the adjusted R-Square, which adjusts R-Square for how many variables there are in the equation. R-Square ranges from 0 to 1; the closer to one (1.0) R^2 is, the better fit the "line" is to the data. In this example, when all six variables are included, R-Square is 0.9955 and the adjusted R-Square is 0.9925, indicating a good fit.

Step 5: Click on Graph to view the residual plots. It is a good idea to view plots of residuals. The plots are helpful to determine if regression analysis is appropriate. A pattern other than a random horizontal band about zero indicates that the assumptions necessary for a regression procedure may be violated. You have options of producing plots of the residuals, and/or predicting values for the dependent variable based on values of the independent variable(s).

Correlation Analysis
(Analyze/Regression and Correlation/Correlation, Pearson & Spearman)

The correlation coefficient is a measure of the strength of the linear relationship between two variables. WINKS calculates both Pearson's and Spearman's (rank) correlation coefficients.

Like regression analysis, correlation assumes that the relationship between the two variables is linear. That is, when one of the variables is plotted against the other, the data points should show a straight line pattern (no curves). Unlike regression, correlation assumes that both variables are independent; neither is dependent on, causes or influences the other. You may wish to create a scatterplot of the data to check for linearity. The correlation coefficient takes on values between -1 and 1, with values close to -1 or 1 indicating a strong relationship between the two variables. A value close to 0 indicates a weak or non-existent relationship. A negative value shows a negative or inverse relationship--as one variable increases, the other decreases. A positive value shows a positive or direct relationship--as one variable increases, the other also increases. Pearson's correlation coefficient (Pearson's r) assumes that both populations are well approximated by a normal distribution, and that their joint distribution is bivariate normal. WINKS calculates Pearson's r and R^2 (the coefficient of determination equal to r squared), and performs a t-test on the significance of rho (the population correlation coefficient) and reports a p-value. This t-test is a test of the hypotheses:

Ho: rho = 0
Ha: rho <> 0

A low p-value (less than 0.05, for example) is usually taken to indicate that the correlation is significant (Ho is rejected).

WINKS also calculates and reports Spearman's rank correlation coefficient. This result does not assume normality and is based on the ranks of the data rather than the data values themselves. Spearman's r (r_s) is calculated by ranking the data within each of the two groups, then finding the Pearson correlation for the rank data. Thus, r_s measures the linear relationship between the ranked data and thus measures the monotonic relationship between the original variables, i.e., does the variable increase or decrease consistently as the other values increased. Spearman's rank correlation coefficient falls between -1 and 1, like Pearson's r and is interpreted similarly.

You may be interested in measuring the strength of the linear association between the numbers of registered handguns and the number of homicides. Both variables may well be influenced by other factors. You want to know if they are related.

Step 1: Open the database named HANDGUNS.

Step 2: From the Analyze menu, select "Regression and Correlation" and "Correlation - Pearson and Spearman."

Step 3: Select HAND_REG and HOM_RATE as the two variables to analyze and click Ok. The results are:

```
Variables used: HOM_RATE and HAND_REG

Number of cases used: 13

Pearson's r (Correlations Coefficient) = 0.7263 R-Square = 0.5275

Test of hypothesis to determine significance of relationship:
H(null): Slope = 0 or H(null): r = 0

(Pearson's) t = 3.504481 with 11 d.f. p = 0.005
(A low p-value implies that the slope does not = 0.)

Spearman's Rank Correlation Coefficient = 0.7527
```

In this example, Pearson's r is 0.7263 and R^2 is 0.5275. The t-test of significance of the relationship has a low p-value 0.005, indicating that the correlation is significantly different from zero. Spearman's rank correlation coefficient is 0.7527. The investigation must determine whether or not these correlations are large enough to be important. How substantial a correlation of 0.7263 is depends on the specific situation and the judgment of the researcher. To check whether the two variables are linearly related, you may wish to produce a scatterplot. To do so, you can use the "Graphical Correlation Matrix" or "Simple Linear Regression" option, or you can use the XY plot in the Graphs menu.

Correlation Matrix
(Analyze/Regression and Correlation/Correlation Matrix)

To display the correlation matrix of the Longley data:

Step 1: Open the database named LONGLEY.DBF.

Step 2: From the Analyze menu, select "Regression and Correlation" then choose the "Correlation Matrix" option.

Step 3: Select all the fields. WINKS performs the calculations and displays the 7 by 7 matrix of correlations:

	DEFLAT	GNP	UNEMP	ARMED	POP	TIME	TOTAL
DEFLATOR		.992	.621	.465	.979	.991	.971
		(0.0)	(0.01)	(0.07)	(0.0)	(0.0)	(0.0)
		[16]	[16]	[16]	[16]	[16]	[16]
GNP			.604	.446	.991	.995	.984
			(.013)	(.083)	(0.0)	(0.0)	(0.0)
			[16]	[16]	[16]	[16]	[16]
UNEMP				-.177	.687	.668	.502
				(.511)	(.003)	(.005)	(.047)
				[16]	[16]	[16]	[16]
ARMED					.364	.417	.457
					(.165)	(.108)	(.075)
					[16]	[16]	[16]
POP						.994	.96
						(0.0)	(0.0)
						[16]	[16]
TIME							.971
							(0.0)
							[16]

Key: Correlation
(p-value)
[count]

Only half of the array is displayed since the other half is a mirror image. The diagonal entries are also omitted since they are all one; a variable is always perfectly correlated with itself. Each entry in the array consists of three numbers. The first (upper) is the Pearson's correlation coefficient for the two (row and column) variables of that entry. The second (middle) number, in parentheses, is the p-value of the t-test for Ho: rho = 0 vs. Ha: rho <> 0. The third (bottom) number, in brackets, is the sample size, or number of paired observations used in the calculations.

Both the correlation coefficient and the p-value are interpreted as they are for any correlation of two variables. In this array, for example, POP and TIME are highly correlated (r=0.994, p=0.00) but POP and ARMED are not (r= 0.364, p=0.17). Notice that ARMED and UNEMP have a negative correlation (r=-0.177); as one increases the other decreases. However, since p is large (0.51), we cannot conclude that this correlation is significantly different from zero. Care must be taken when running a multitude of tests at a given significance level. As the number of tests increases, the chances of finding a significant relationship when none really exists increases.

Graphical correlation matrix
(Analyze/Regression and Correlation/Graphical Correlation Matrix)

You may display an array of scatterplots (XY plots) to see in one screen relationship of up to ten variables. To perform this analysis on the LONGLEY data, follow these steps:

Step 1: Open the database named LONGLEY.

Step 2: From the Analyze menu select "Regression and Correlation" then choose the "Graphical Correlation Matrix" option.

Step 3: Select all the fields for this analysis. WINKS will perform the calculations and display the matrix of scatterplots.

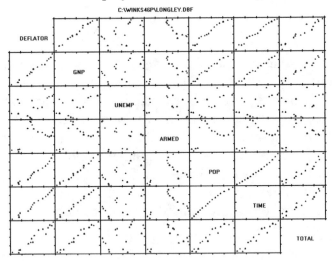

These scatterplots are a visual way of examining the relationships between pairs of variables. It allows you to determine if a relationship exists between the variables, and allows you to see if that relationship is linear. The more highly correlated two variables are, the more tightly clustered about a straight line are the points on the scatterplot. In this case you can see that GNP looks highly correlated with TIME (we know r= 0.995 from a previous example) whereas GNP and UNEMP do not look as related , although still statistically different from 0 (r = 0.604). Notice also that the relationship between ARMED and TIME looks related, but not in a linear fashion (r = 0.417). You can use this graphical correlation matrix to examine the relationships between variables before using them in a multiple regression analysis.

Point Bi-serial Correlation

(Analyze/Regression and Correlation/Point Biserial Correlation)

The point bi-serial correlation is used when one of the measures is continuous and one is dichotomous (0,1). For example, using the database on disk named POINTBS.DBF, select SCORE as the "X" (first) field and YESNO as the "Y" (0,1 type) field. The results are (edited):

```
Dependent variable =SCORE

Independent variable = YESNO

Point bi-serial correlation = 0.5948

t = 6.535 with 78 degrees of freedom. p <= 0.001
```

When the p-value is small (less than 0.05), you can conclude that there is evidence to reject the null hypothesis and to support the alternative hypothesis. For this example, since p is small, you can conclude that there is a significant relationship between SCORE and YESNO. You might note that the t-test in this case is equivalent to an independent group t-test using the 0,1 variable as the grouping variable.

Analysis of Count Data – Frequencies and Crosstabulations

Frequencies Analysis
(Analyze/Crosstabs, Frequencies, Chi-Square/Frequencies)

The Frequencies analysis "counts" the occurrence of each data value for a variable and displays that information in a table. Suppose you want to know the number of people in each STATUS group in the EXAMPLE data set.

Step 1: Open the database named EXAMPLE. From the Analyze menu, choose "Crosstabulations, Frequencies, Chi-Square," then "Frequencies."

Step 2: Select STATUS and click "Add". Select GROUP as the grouping field. Click Ok. The following information is displayed:

```
------ GROUP = A

                      Frequency Table for STATUS

                                    Cumulative   Cumulative
        STATUS      Frequency   Percent   Frequency   Percent
        -----------------------------------------------------
          2            1        9.09         1         9.09
          3            3       27.27         4        36.36
          4            1        9.09         5        45.45
          5            6       54.55        11       100.0

------ GROUP = B

                      Frequency Table for STATUS

                                    Cumulative   Cumulative
        STATUS      Frequency   Percent   Frequency   Percent
        -----------------------------------------------------
          1            2        6.9          2         6.9
          2            6       20.69         8        27.59
          3            2        6.9         10        34.48
          4            4       13.79        14        48.28
          5           15       51.72        29       100.0
```

etc...

Goodness-of-Fit Analysis
(Analyze/Crosstabs, Frequencies, Chi-Square/Goodness-of-fit)

A goodness-of-fit test of a single population is a test to determine if the distribution of observed frequencies in the sample data closely matches the expected number of occurrences under a hypothetical distribution of the population. The data observations must be independent and each data value can be counted in one and only one category. It is also assumed that the number of observations is fixed. The hypotheses being tested are

Ho: The population follows the hypothesized distribution.
Ha: The population does not follow the hypothesized distribution.

A Chi-Square statistic is calculation and a decision can be made based on the p-value associated with that statistic. A low p-value indicates rejection of the null hypothesis. That is, a low p-value indicates that the data do not follow the hypothesized, or theoretical, distribution.

For example, data for this test comes from Zar (1974), page 46. According to a genetic theory, crossbred pea plants show a 9:3:3:1 ratio of yellow smooth, yellow wrinkled, green smooth, green wrinkled offspring. Out of 250 plants, under the theoretical ratio (distribution) of 9:3:3:1, you would expect about

(9/16)x250=140.625   yellow smooth peas,
(3/16)x250=46.875   yellow wrinkled peas
(3/16)x250=46.875   green smooth peas
(1/16)x250=15.625   green wrinkled peas

After growing 250 of these pea plants, you observe that

152 have yellow smooth peas
 39 have yellow wrinkled peas
 53 have green smooth peas
  6 have green wrinkled peas

To perform this analysis, use the following steps:

Step 1: From the Analyze menu, select "Crosstabulations, Frequencies, Chi-Square" then choose the "Goodness-of-Fit" option.

Step 2: You will be prompted to enter the number of categories. In this case, enter 4 for the four categories of peas (yellow smooth, yellow wrinkled, green smooth, green wrinkled).

Step 3: A dialog box will appear allowing you to enter the observed data and either the expected values or ratios. In this example, click (select) on the check box labeled "Check to enter ratios rather than expected values." Enter the observed values from the table above:

152, 39, 53, 6

and the ratios

9,3,3,1

Press Tab to move from cell to cell. Click Calculate and Exit and the following information is displayed:

```
           Obs.      Exp.
        |-------|-----------|
1       |   152|    140.625|
        |-------|-----------|
2       |    39|     46.875|
        |-------|-----------|
3       |    53|     46.875|
        |-------|-----------|
4       |     6|     15.625|
        |-------|-----------|
            250        250.
```

Calculated CHI-SQUARE = 8.97
Degrees of freedom = 3 Appx p = 0.031

At a 0.05 level of significance, this p-value indicates that there is enough evidence to reject the null hypothesis that the observed values follow the theoretical distribution. That is, the test (at the 0.05 significance level) suggests that a 9:3:3:1 ratio of yellow smooth:yellow wrinkled:green smooth:green wrinkled peas is not an appropriate distribution for the population from which these data are taken.

Note: you can perform several analyses in the Goodness-of-Fit dialog box, and each will be displayed in the viewer when you select End.

Crosstabulation Analysis (Chi-square)

Crosstabulations can be used to perform a chi-square test for independence or a chi-square test for homogeneity. A two-way table is constructed that displays the number of counts for each category. It must be possible to assume that the data observations are independent and that each data value can be counted in one and only one category. It is also assumed that the number of observations is fixed. WINKS allows you to enter data for a two-way table from the keyboard or from a database.

Entering Data from the Keyboard
(Analyze/Crosstabs, Frequencies, Chi-Square/Chi-Square from Keyboard)

When you choose to enter the two-way table from the keyboard WINKS will ask you the size of the table (number of rows and columns). A blank table will be presented on the screen, and you will then be prompted to enter a number in each cell of the table.

Entering Data from a Database
(Analyze/Crosstabs, Frequencies, Chi-Square/Crosstabulations/Chi-Square)

If you choose to enter the information from a database, you will be prompted to indicate what tables are to be calculated. Select one or more fields for the "Data field" (top right hand list box) and select one or more fields for the "By Var" field (bottom right hand side list box). For example if you select fields A and B in the first box, and C in the second box, the tables A x C and B x C will be calculated. See database example below.

For all tables, you will be prompted to specify what output options you want included in the output tables:

- Frequencies only
- Include Expected Values
- Include Expected Values and Percents
- Include Expected Values, Chi-Contribution and Percents
- Include Percents
- Include Expected Values and Chi-Contribution

Percents display row, column and total percent of the number in each cell of the table. The contribution to Chi-Square shows how much a particular cell contributed to the size of the Chi-Square statistic. This often comes in handy when you are trying to discover what may have caused a table to result in a low p-value (a high Chi-Square statistic).

For a test for independence, a contingency table looks at two categorical variables from a single sample of one population and tests whether the two variables are related in some way, (e.g., are sex and hair color related?) The hypotheses being tested are:

Ho: The variables are independent of each other. (There is no association between them).
Ha: The variables are not independent of each other.

A Chi-Square statistic is calculated, with (r-1)(c-1) degrees of freedom where r is the number of rows and c the number of columns. A low p-value indicates rejection of the null hypothesis.

WINKS reports both the chi-square statistic and the p-value. If the expected value in one or more cells is less than 5, the chi-square test may not be valid. A warning to this effect appears on the screen if appropriate. In the case of a 2 by 2 table, Fisher's Exact Test and the chi-square with Yates' correction are also performed and results displayed. Note: Tables as large as 15 columns by 100 rows may be created by reading data from a database. If there are more categories than this, WINKS combines remaining categories in a group called REST. To prevent his, you might combine some groups.

Data for this example are observations of the number of beetles and bugs on the upper and lower sides of leaves (Zar,1974, page 292).

2 by 2 Contingency Table Data

	Beetles	**Bugs**
Upper Leaf	12	7
Lower Leaf	2	8

To perform this analysis, follow these steps:

Step 1: From the Analyze menu, select "Crosstabulations, Frequencies, Chi-Square" and choose "Crosstabulations, Chi-Square - From Keyboard" option.

Step 2: You will be prompted to give the size of the table. When asked for the number of rows and columns, type 2, 2 and press Enter. You will then be prompted to select output options. For this example, just select Frequencies. An empty table will appear.

Enter the counts for each category into the appropriate cell, and
choose Calculate. Preliminary results will appear on the status bar
a the bottom of the screen. You can perform calculations on
several tables, and all results will appear in the viewer when you
select Exit.

```
2-Way Contingency Table

FREQUENCY|    |    |  TOTAL
         ----------------------------
         | 12|   7|   19
         ----------------------------
         |  2|   8|   10
         ----------------------------
  TOTAL    14   15    29
          48.3 51.7 100.0
```

WARNING - Some Expected values less than 5. Chi-
Square may not be valid.

```
Statistic                   DF      Value    p-value
---------------------------------------------------------------
Chi-Square                    1      4.887    0.028
Yates' Chi-Square   1        3.312   0.069
Fisher's Exact Test      (one-tail)           0.033
                         (two-tail)           0.111
Phi Coefficient             .411
Cramer's V .411
Contingency Coefficient     .380
Relative Risk                       3.158
Odds Ratio                          6.857
Sensitivity                          .857
Specificity                          .533
```

Sensitivity and Specificity calculations are based on a
table where the cells are in the following pattern:
TP FP
FN TN
T=True, F=False, P=Positive, N=Negative

Step 3: The calculated chi-square statistic in reported as 4.89 with
a p-value of 0.028. The chi-square with Yates correction is 3.31
with a p-value of 0.069 and the Fisher Exact Test (two-tailed) has a
p-value of 0.050. Because one of the cells produces an expected
value less than 5, WINKS gives a warning that the chi-square

analysis for this data may not be valid. Given this warning, it is best to rely on the Fisher's Exact Test for making a decision.

A low p-value indicates rejection of the null hypothesis. At a 0.05 significance level, the Fisher's Exact Test p-value of 0.050 indicates (on the borderline) that there is enough evidence to reject the null hypothesis of independence of the two variables and to conclude that leaf side and type of insect are not independent. In this case it appears that beetles prefer the upper sides of leaves and bugs are about split in their preference. In the case of the Yates results, this decision is marginal.

Notes on 2x2 Table Statistics

RELATIVE RISK is given by the formula

```
         a/(a + b)
RR  =  -----------
         c/(c + d)
```

where the two by two table is

```
              Factor 1
                +    -
              ---------
              +|a  |  b|
Factor 2       ---------
              -|c  |  d|
              ---------
```

ODDS RATIO is calculated by

```
        a / b
OR  =  -----
        c / d
```

SENSITIVITY is calculated by

```
        a
SEN = -----
      a + c
```

Sensitivity is a measure of the ability to call positive those patients/subjects that have the disease/condition.

SPECIFICITY is calculated by

```
       d
SP  = -----
      b + d
```

Note: Sensitivity and Specificity calculations are only available on the crosstabulations with data entered from the keyboard. Results are based on a table where the cells are in the following pattern:

```
TP  |  FP
_____
FN  |  TN
```

Where T=True, F=False, P=Positive, and N=Negative.

Specificity is a measure of the ability to call negative patients/subjects that do not have the disease or condition. For more information on these four statistics, reference any biostatistics or epidemiology text. One example is *Basic Biostatistics in Medicine and Epidemiology*, A. A. Rimm, Appleton-Century-Crofts, 1980.

Example: Crosstabulation– Homogeneity Hypothesis
(Analyze/Crosstabs, Frequencies, Chi-Square/Crosstabulations/Chi-Square)

A crosstabulations analysis may also be used to perform a chi-square test for homogeneity. In this case the two-way table looks at variables in samples from two populations and tests whether the variables follow the same distribution for both populations, that is, whether the populations are homogeneous.

The test statistic and procedure are the same as for the chi-square test for independence. The data are organized so that the populations being tested for homogeneity are represented as values (groups) of a grouping variable. The hypotheses being tested in this case are:

Ho: The populations are homogeneous.
Ha: The populations are not homogeneous.

To use the crosstabulations procedure to test for homogeneity, let the rows of the table represent the categories of the variable and the columns the different populations. (Of course, for test purposes it doesn't matter whether the rows or columns represent the populations.)

If entering the data from the keyboard, simply enter the totals for each category. If creating a database, you need a field for the categorical variable you wish to use in the test for homogeneity, and a grouping variable which identifies the population from which each observation comes. Each record represents one observation. There may be other fields in the database for other variables in these same populations and you can do separate crosstabulation analyses on them.

Suppose you want to check whether the ratio of men and women is the same in three different departments of a company. You obtain the following data:

	Dept 1	Dept 2	Dept 3
Men	10	45	15
Women	8	22	4

The calculated chi-square statistic in this case is 2.30 with a p-value of 0.317. A decision can be made using this p-value of the test. A low p-value (less than the chosen significance level) is usually taken to indicate rejection of the null hypothesis. At a 0.05 significance level, the p-value of 0.317 indicates that there is not enough evidence to reject the null hypothesis of homogeneity of the three departments. That is, you cannot conclude that the departments are significantly different with respect to sex of employees based on this test.

McNemar's Test
(Analyze/Crosstabs, Frequencies, Chi-Square/McNemar's Test)

McNemar's test is appropriate for use with paired, dichotomous data. This test is sometimes called a test for related samples or a test for the significance of changes. It is useful for comparing paired or related observations in which the response is dichotomous, that is, the response is one of only two possible outcomes. McNemar's test is the 2 by 2 version of Cochran's Q test described in the section on non-parametric tests. The test assumes that any pair of observations is independent of any other pair of observations, although clearly the observations within a pair are not independent of each other.

For example, you may wish to know if a certain advertisement has an effect on the impression consumers have of a product. You could select a group of people and check their impressions of the product before viewing the advertisement, and again after viewing the ad. You record the reactions of each person as "favorable" or "unfavorable" both before and after seeing the ad. Thus, there are four categories of before-after responses:

"favorable-favorable",
"favorable-unfavorable",
"unfavorable-favorable",
"unfavorable-unfavorable"

You want to know if there is a difference in the reactions before and after viewing the advertisement. It is possible to use McNemar's test if you have only the totals for each of the four categories, or if you have the record of each individual response. The hypotheses being tested are:

Ho: The proportions (of "favorable" and "unfavorable" reactions) in the two groups (before and after) are the same.
Ha: The proportions in the two groups are not the same.

In other words, you are testing whether there is a change in the number of people who react favorably to the product after seeing the advertisement. The test statistic used is:

$$Q = (B - C)^2/(B + C)$$

where B and C are the number of "favorable-unfavorable" and "unfavorable-favorable" reactions. This test statistic approaches a chi-square distribution with one degree of freedom. The chi-square statistic is useful if the number of pairs of observations is at least 10.

WINKS displays both the chi-square statistic and the p-value of the test. The p-value associated with this test can be used to make a decision. A small p-value (less than the chosen significance level, e.g., 0.05) is usually taken to indicate rejection of the null hypothesis.

For example, in the test of the effect of an advertisement, suppose 20 people participated with the following results (listed on the next page), where 1 is the code for "favorable" and 0 for "unfavorable". For McNemar's test, the data must be coded as 0 or 1, representing the two possible responses. To perform this analysis, follow these steps:

Step 1: Open the database named MCNEMAR.DBF. If you choose to create the database yourself, you can choose the New Database option from the File menu, then use the pre-defined database structure option named "For paired t-tests or McNemar's Test."

Step 2: From the Analyze menu, select "Crosstabulations, Frequencies and Chi-Square" then choose "McNemar's test."

You will be prompted to choose the two fields (groups) you wish to compare. In this case, there are only two fields. Select BEFORE and AFTER. Then select which options you want to appear on the output table. WINKS will perform the calculations and display the results in the viewer.

The chi-square statistic in this case is 0.57, and the p-value is 0.450. The p-value is large, so the null hypothesis of equal proportions is not rejected. That is, there is not enough evidence to say that there is a difference in the reactions before and after viewing the advertisement.

Comparison of Proportions
(Analyze/Crosstabs, Frequencies, Chi-Square/Proportions Comparison)

To compare proportions select Proportion Comparison in the Crosstabs menu. You will be prompted to enter two proportions (or two observed counts) and the sample size associated with each proportion. For example, for the first proportion enter 10 observations out of 20 and for the second enter 15 observations out of 75. When you select Calculate then Exit, the following results are displayed:

```
Proportion(1) = 0.5    Z = 2.707
Proportion(2) = 0.2    P = 0.007 (two tail)
```

This tells you that the z-statistic for the comparison of these proportions is 2.707, and the p-value associated with the test is 0.007. Thus, there is a significant difference in these proportions.

Life Table and Survival Analysis

Survival Analysis is used to Analyze the survival experience of a group of persons or components. In medical research, survival analysis is helpful is studying the survival of patients under one or more conditions. In industry, the survival may be that of a component such as an electronic switch or a gear.

To perform a survival analysis, data must be in the following form:

1) a **TIME variable** which contains a time (e.g., minutes, days, years, etc.) in which the subject or component has been observed to be alive (not failed).

2) a **CENSOR variable** which must take on the values 0 or 1, where 1 means the subject has died (failed), and a 0 means the subject was still alive (not failed) at the last available time period.

3) optionally, a **GROUPING variable** which may have up to ten values (numeric or character), i.e., the data may be in groups.

You can choose from two types of life tables, Actuarial or Kaplan-Meier. The Actuarial method uses fixed length intervals in the table, and the Kaplan-Meier table uses intervals based on the data. A life table for each group is produced which includes, for each time interval, the number entered, withdrawn, lost, dead, exposed, the proportion dead, proportion surviving, cumulative proportion surviving, and other information.

A plot is given for the cumulative proportion surviving in the group(s) against time. If more than one group is entered, a Mantel-

Haenszel log-rank test is performed to test the hypothesis of equal survival patterns for the groups. A reference to how this test is developed is covered in Matthews and Farewell (1988).

Actuarial Life Table Analysis
(Analyze/Life Tables & Survival Analysis/Actuarial Life Table)

The data for this example are in the LIFE.DBF database on the WINKS disk. These data are from Prentice (1973). To perform this analysis, follow these steps:

Step 1: Open the database named LIFE.DBF.

Step 2: From the Analyze menu, choose "Life Tables and Survival Analysis" then select "Actuarial Life Table Analysis".

Step 3: Select SURVIVAL and CENSOR fields (in that order), then select

GROUP as the Group field. A portion of the LIFE database is shown here:

SURVIVAL	CENSOR	GROUP
72	1	1
411	1	1
228	1	1
11	1	1
25	0	1
etc...		

The first column is the SURVIVAL field with entries of length of life, or length of survival. The second column is the CENSOR field, an indicator of whether the subject has failed (died) or not at the last observed time period. 1 means failed, 0 means not failed (still alive). The third column contains a grouping variable. In this case it is either 1 or 2. Group 1 may represent one treatment, while group 2 represents another kind of treatment. The objective is to compute survival curves to see if the treatments provide different average survival distributions.

Step 4: Specify a desired interval length or you can use the default length by simply pressing Enter. For this example, press Enter to select the default length.

Step 5: WINKS will perform the calculations and display the results in the viewer.

```
Actuarial Table Analysis

SURVIVAL GROUP:0
```

Interval	Enter Alive	With- drawn	Dead	Exposed	Prop. Dead	Alive	Cumulative Survival	S.E.
0.0 99.0	38	2	22	37.00	0.5946	0.405	1.0000	
0.0000								
99.0 198.0	14	0	4	14.00	0.2857	0.714	0.4054	
0.0807								
198.0 297.0	10	1	3	9.50	0.3158	0.684	0.2896	
0.0756								
297.0 396.0	6	0	2	6.00	0.3333	0.667	0.1981	
0.0677								
etc...								

Interval	Exposed	Cumulative Survival	Lower Bound	95% confidence limits Upper Bound	Hazard Function
0.0 99.0	37.0	1.000	1.000	1.000	0.009
99.0 198.0	14.0	0.405	0.247	0.564	0.003
198.0 297.0	9.5	0.290	0.141	0.438	0.004
etc...					

```
Summary Table

     Total    Dead   Withdrawn   Percent Censored
-------------------------------------------------
       38      35        3           7.89%
etc...

(Continues for Group 1)

Mantel-Haenszel comparison of survival curves
-------------------------------------------------
Chi-Square = .7472     with 1 D.F    appx p = 0.388
```

The first table includes the numbers of subjects entered alive, withdrawn, dead, exposed, the proportion dead, proportion alive, cumulative survival proportion and standard error for the first group. The second table includes 95% confidence limits on the cumulative survival proportion .

From the table, you can see that, in the first group, 22 of 37 exposed, or 59.5% died in the first interval (0.0-99.0) and two were withdrawn. In the second group, 12 of 23.5 exposed (51.1%) died and one was withdrawn in the first interval. At the end of the report, WINKS reports the results of the Mantel-Haenszel comparison of the two curves. The hypotheses being tested are:

Ho: The survival curves are the same.
Ha: The survival curves are not the same.

In this example, the Mantel-Haenszel comparison procedure results in a chi-square statistic of 0.7191 and a p-value of 0.397. This p-value is much too large to reject the hypothesis of equal curves. This indicates that the two distributions are not statistically significantly different - thus neither treatment is superior in terms of survival distributions.

Step 6: Click on Graph to display a graph of survival curves.

Kaplan-Meier Analysis
(Analyze/Life Tables & Survival Analysis/Kaplan-Meier Analysis)

Data for a Kaplan-Meier analysis is the same as for the Actuarial analysis. Follow the same steps listed in the above example, except you will not enter an interval length.

The Kaplan-Meier life table contains most of the same information as the Actuarial Life Table. However, instead of the time intervals being fixed, the time intervals are based on time values from the data. The Mantel-Haenzel statistic will be the same for both Life Table analysis types.

On the Kaplan-Meier survival plot, you may optionally choose to include markers indicating censored values. The following plot was created using the LIFE.DBF data set:

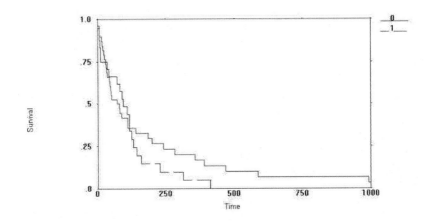

Analyze From Summary Data

This menu option on the Analyze menu repeats several analysis
options found elsewhere in the menu system. See explanations in
the sections pertaining to that analysis type.

Simulations and Demonstrations

This menu item on the Analyze menu is used to describe several
statistical concepts. They include:

- 95% Confidence Interval Simulation
- Flip a coin demonstration
- Demonstrate Central Limit Theorem

This ends the section for WINKS BASIC. The next section
includes features for WINKS Professional.

WINKS

PROFESSIONAL EDITION FEATURES

Part 5 – Professional Edition Procedures

Advanced Tabulation Procedures

The WINKS Professional Edition provides the following advanced tabulation procedures:

- Advanced Tabulation – Create tables of counts, means, etc with up to three variables.
- Mantel Haenszel Comparison – Used to analyze multiple 2x2 contingency tables
- Inter-Rater Reliability/KAPPA – Used to analyze agreement between raters

Advanced Tabulation (Analyze/Advanced Tabulation)

The WINKS Advanced Tabulation module allows you to create a wide variety of tables containing counts, averages, sums and more. This procedure can be used to create reports about your data quickly, easily, and in a convenient format. Tabulate categorical data up to two rows and two columns, and up to 128 categories in each row or column. Plus, you can report counts, averages, sums or standard deviations in each cell of your table. You can also request subtotals and totals by row or columns and output tables using a "by" variable. Unlike the Crosstabulation procedure in the BASIC edition of WINKS, which is used for Chi-Square analysis, the Tabulation procedure is primarily a reporting and data reduction program.

What are Tabulation Variables?

Your database for the tabulation procedure must contain at least one categorical variable. Usually, a tabulation database will contain a number of categorical variables and one or more continuous variables. The tabulation procedure cannot use date or logical variables.

Note: You may want to use the data editor to manipulate your data to specify groups or categories that you wish to use in the tabulation procedure. The new Recode procedure described later in this manual can help you do this.

Selecting Variables for Use in Tabulation

Once you open a database and select "Advanced Tabulation" from the *Analyze* menu, a dialog box will appear allowing you to specify which variables you want to be available for the Tabulation procedure. Select all variables you may want to use since you will be given a chance within the Tabulation procedure to create any number of tables using only a few or all of the selected variables.

Understanding Missing values in Tabulation: When you make the initial selection of which variables to use in the Tabulation procedure, notice the following radio button at the bottom of the selection dialog box:

O   Include records with missing values

If this option is NOT selected, all records with missing values in any of the fields you have selected will be excluded from all tables you create. If you SELECT this item, Tabulation will pass all records to the Tabulation procedure, and it will only eliminate records with missing values as you are building your tables, which means that if a record contains a missing value in one field, but you do not use that field in creating a table, the information in that variable will be used. Once you have selected which variables to use in Tabulation, the tabulation main dialog box will appear, as shown below.

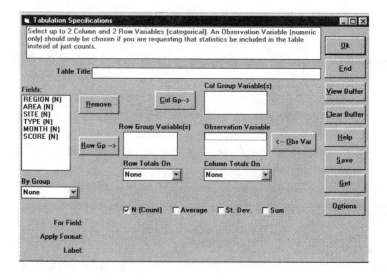

In the Tabulation dialog box, you specify how you want your table created using these options:

Fields: The Fields list box contains all of the field names you selected to be available for the Tabulation procedure.

Row Gp and Col Gp: These buttons are used to tell the procedure which fields to use to create Columns and Rows in the table. For best results, select only fields containing categorical variables. The variables may be either numeric or character type. You may select up to two columns and two rows.

Remove: If you place a variable in the Col Gp or Row Gp boxes, then decide differently, highlight the name of the field in the Col or Row box and click the Remove button.

Obs Var: This button allows you to specify an observation variable to be used in the table. This variable should be of a numeric type and be continuous. Its purpose is to allow you to create tables containing averages, sums or standard deviations. For example, if you create a table for stores by state, the observation may be sales. This would give you a table that shows sales within each store in each state.

Table Title: This text box allows you to specify a title that will appear at the top of the output.

By Group Combo Box: This selection box allows you to optionally specify a "by" group variable. The field should be categorical, and may be either numeric or character. When selected, this will cause tables to be subgrouped on this variable. For example, if you have a "Month" variable and create a store by state table, the program will create a new table for each month.

Row Totals On and Column Total On: These option boxes allow you to choose to have your tables include totals on variables selected for rows or columns. You must have selected an Obs (Observation) variable for these options to work.

Include Sub Totals: These boxes may be optionally selected if matching Row or Col Totals options are also selected. They cause the program to include subtotals within the tables.

N, Average, St. Dev, Sums: Normally, a table will report the count (N) of observations that fall within each cell. Using these options, you can specify which specific statistics will appear in the table. If you select Average, St. Dev or Sum, you must also have selected an Observation variable.

Field Format and Labels Options: At the bottom of the
Tabulation dialog box, you may select formats and labels for the
variables selected in Row Gp, Col Gp, and By Group. Under each
field name is a drop down box that lists all defined field formats
(see "Creating Field Formats" later in this section.) Select the field
format that fits the field name. For example, if your field is
Gender, where the data contain 0=Male and 1=Female, you would
select the field format called SEX01. This will cause the
information in the table to be reported as Male and Female instead
of the more cryptic 0 and 1.

The Label option allows you to select a label to be used in place of
the field name. For example, if your field is called MTH, you may
want to label it "Month." Then, instead of MTH appearing in the
table, the label Month will be used.

The Ok Button: Once you have defined your table, click on the
Ok button to tell the program to create the table. Depending on the
size of your database, this usually takes only a second or two. You
may then view the table created by clicking on the View Buffer
Button.

End Button: This button ends the Tabulation procedure and
returns you to the main WINKS menu. Any tables you have
created will be displayed in the WINKS Viewer.

View Buffer: You may see the tables you have created in the
Tabulation procedure without having to return to the main WINKS
menu by clicking on the View Buffer button. From the viewer, you
may print, copy, or save the information. When you exit the
procedure, you return to the Tabulation dialog box. From there you
can create more tables. Each time you create a table, it is appended
onto the information already in the buffer.

Clear Buffer: If you create a table you do not want to keep, click
on the Clear Buffer button to clear information in the buffer. This

enables you to experiment with the options in the Tabulation dialog box to get your table just right before printing or saving it.

Help: Displays information about using the Tabulation Procedure.

Save: After creating a table definition that you like, you may want to save all of your choices to use them again. (Perhaps you will add new information to your database.) When you click on the Save button, you will be given an opportunity to enter a save file name (must be 8 characters or less.)

Get: The Get button allows you to get a previously saved table definition. You may use this definition on data from a different database than the one used to create it as long as the same variables previously used are available in the different database.

Options: When you click on this button, an Options dialog box is displayed, as shown in the figure below.

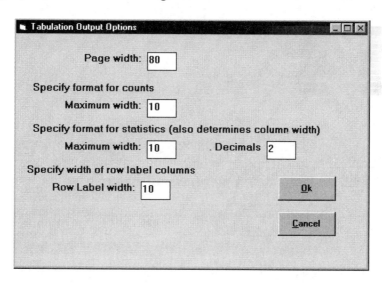

In the Options dialog box you may select the width of the printed page, the formatting of numbers in the table and the Row label width. All of these options are saved when you select the Save button. For example, if you are creating a wide table to be printed in landscape mode you may want to change the Maximum width from 80 to 130.

Tabulation Program Features: When creating a table, you should keep this information in mind.

- If your table is wider than the Maximum Width, the table will be created in sections.
- If you save your table and import it or copy it into a word processor, remember to display it using a monospaced font such as Courier. If you display the table in a font such as Times Roman, the columns will not line up properly.

Tabulation Example

Follow this example to create an example table.

Step 1: Begin the WINKS program and open the database named TABDATA.DBF. From the Analyze menu, select Advanced Tabulation.

Step 2: Select the fields that will be available in the tabulation procedure. Highlight all of the field names and click Add, then click Ok. The Tabulation dialog box will appear.

Step 3: Select the Region field for the Col variable, and the Area field as the Row variable.

Step 4: On the By Group drop down dialog box, select the Month variable as the "By" Variable.

Step 5: Apply formats to the selected fields. For the By Field, select MTH as the format for MONTH. Select the AREA format for AREA and the REGION format for REGION.

Step 6: Enter labels for the variables: Enter the label Month for MONTH, Area for AREA and Region for REGION.

Step 7: Click Ok to create the table, then click on View Buffer to view the resulting table. Exit the viewer to return to the Tabulation dialog box.

Step 8: Create other tables or End the Tabulation procedure to return to the WINKS main menu. The tables resulting from this example are shown below:

```
---------- Month = Jan ----------

-----------------------------------
|         |         Region        |
|         |---------------------- |
|         |  North  |   South     |
|---------|---------|-------------|
| Area    |         |             |
|---------|         |             |
|Inside   |       8 |           3 |
|---------+---------+-------------+
|Outside  |       8 |           8 |
|---------+---------+-------------+

---------- Month = Feb ----------

-----------------------------------
|         |         Region        |
|         |---------------------- |
|         |  North  |   South     |
|---------|---------|-------------|
| Area    |         |             |
|---------|         |             |
|Inside   |       8 |           3 |
|---------+---------+-------------+
|Outside  |       8 |          10 |
|---------+---------+-------------+

Key: Statistics in table are: N
```

Notice that two tables were created, labeled Jan and Feb. This resulted from the MONTH "By" variable (the database contains

information on two months.) The numbers in the table are the
counts of records containing information for Area and Region.

Experiment with this table by adding the Score Field as the
Observation variable, and select to display average, totals, and
subtotals.

Defining Format Definitions for Tables

In the previous example, several formats were used to enhance the
look of the table. This allowed your output to report the name of
the month (i.e., Jan, Feb) instead of a more cryptic 1 and 2. The
tabulation procedures come with a set of automatically defined
formats, but you can also create your own.

The file called FORMAT.DAT, which will be in your WINKS
directory, contains format definitions. You can edit this file with
any ASCII text editor including Word, WordPerfect. *(Just
remember to save the results in text format.)* Initially, the format
file contains this information:

```
@SEX01
0=Male
1=Female
@SEX12
1=Male
2=Female
@SEXMF$
M=Male
F=Female
@MTH
1=Jan
2=Feb
3=Mar
etc...
    :
@ENDFORMAT
```

You can add your formats using the same syntax shown here. (1)
Begin the format name with an "@" (2) If the format is character,
place a "$" as the last character in the format name. The last line in
the FORMAT.DAT file must be @ENDFORMAT. Once you enter

new formats in this file, they will appear in the format drop-down boxes within the Tabulation procedure.

Mantel Haenszel Comparison
(Analyze./ Advanced Tabulation, and the Mantel-Haenszel Comparisons)

The Mantel-Haenszel procedure allows you to a series of combine 2x2 tables to adjust for a confounding factor when analyzing the relationship between two dichotomous variables. The point is this; combined table could show a significant relationship, but when the data are stratified, the results could be different. Consider the data set collected at Berkeley University to assess whether men were being given preferential treatment over women in admission to graduate programs (Bickel & O'Connell, 1975, Freedman et al., 1991, pp. 16 - 19). The combined data gives this table:

```
                    ACCEPTED
GENDER           NO      YES
-----------+----------------+------
   FEMALE  |    1259     478 |   1727
     MALE  |    1180     686 |   1866
-----------+----------------+------
    Total  |    2439    1154 |   3593
```

This table produces the following results

```
Chi-Square =    38.421      p    < 0.001
Relative Risk = 1.153
Odds Ratio   = 1.564
```

```
The acceptance rate for women is  0.277
The acceptance rate for women is 0.368
```

From this analysis it is evident that men are accepted at a higher rate than women.

However, consider the tables when they are stratified by type of major. These results are:

Department 1

```
FREQUENCY|  N  |  Y  |   TOTAL
-----------------------------
       F  |   8|  17|    25
-----------------------------
       M  | 207| 353|   560
-----------------------------
   TOTAL      215   370     585
             36.8  63.2   100.0
```

Chi-Square = .254   p= 0.615
Odds Ratio = .803   95% C.I. = (.340,   1.892)

Department 2

```
FREQUENCY|  N  |  Y  |   TOTAL
-----------------------------
       F  | 391| 202|   593
-----------------------------
       M  | 205| 120|   325
-----------------------------
   TOTAL      596   322     918
             64.9  35.1   100.0
```

Chi-Square = .754   p = 0.386
Odds Ratio = 1.133 95% C.I. = (.855,   1.502)

Department 3

```
FREQUENCY|  N  |  Y  |   TOTAL
-----------------------------
       F  | 244| 131|   375
-----------------------------
       M  | 279| 138|   417
-----------------------------
   TOTAL      523   269     792
             66.0  34.0   100.0
```

Chi-Square = .298   p= 0.585
Odds Ratio = .921   95% C.I. = (.686,   1.237)

Department 4

```
FREQUENCY|  N  |  Y  |   TOTAL
-----------------------------
       F  | 299|  94|   393
-----------------------------
       M  | 138|  53|   191
-----------------------------
   TOTAL      437   147     584
             74.8  25.2   100.0
```

Chi-Square = 1.001 p= 0.318
Odds Ratio = 1.222 95% C.I. = (.825,   1.809)

```
Department 5

    FREQUENCY|  N  |  Y  |   TOTAL
    ----------------------
         F  |  317|   24|    341
    ----------------------
         M  |  351|   22|    373
    ----------------------
    TOTAL       668    46      714
                93.6   6.4   100.0

Chi-Square = .384  p= 0.536
Odds Ratio = .828  95% C.I. = (.455,    1.506)

MANTEL-HAENSZEL RESULTS
--------------------------------------------------
The Mantel-Haenszel procedure is used to analyze multiple 2x2
tables adjusting for the factor represented by the multiple tables.

Mantel-Haenszel = 0.12 p = 0.723 (two-sided)
Mantel-Haenszel Average Odds Ratio = 1.031
Appx. 95% C.I. for Odds Ratio = 0.87 to 1.221
```

Thus, although the aggregated data shows significance, when looked at by department, using the Mantel-Haenszel method of adjusting for the confounding factor (department) the results do not show any overall difference in acceptance by gender.

To perform this analysis, select Analyze/ Advanced Tabulation, and the Mantel-Haenszel.

Step 1: Select output detail – Choose Frequencies only.

Step 2: Enter number of 2x2 tables to analyze. Enter 5.

Step 3: Enter the table values (from the tables above)

Step 4: Click okay and the results (as shown above) are be displayed.

Inter-Rater Reliability/KAPPA
(Analyze/Advanced Tabulation/InterRater Reliability/KAPPA)

Cohen's Kappa coefficient is a method for assessing the degree of agreement between two raters. The weighted Kappa method is designed to give partial, although not full credit to raters to get "near" the right answer, so it should be used only when the degree of agreement can be quantified.

For example, using an example for Fleiss (1981, p 213), suppose you have 100 subjects rated by two raters on a psychological scale that consists of three categories. The data are given below:

		RATER A			
		Psyc.	Neuro.	Organic	
		1	2	3	
Rater	Psych 1	75	1	4	80
B	Neuro 2	5	4	1	10
	Organic 3	0	0	10	10
		80	5	15	100

Step 1: To perform this analysis in WINKS, select Analyze/Advanced Tabulation/InterRater Reliability/KAPPA. Select "Frequencies Only" for output options.

Step 2: When prompted for number of Rows/Columns, enter 3.

Step 3: Enter the data from the table. You may select to use either the Cicchetti-Allison or Fleiss Cohen weights. The Cicchetti-Allison weights are used by default.

Step 4: Click Calculate and the results will appear in the WINKS viewer:

```
-------------------------------------------------------
   Inter-Rater Reliability/Kappa
-------------------------------------------------------

   FREQUENCY|    1|    2|    3|   TOTAL
   ----------------------------------
          1|   75|    1|    4|    80
   ----------------------------------
          2|    5|    4|    1|    10
   ----------------------------------
          3|    0|    0|   10|    10
   ----------------------------------
   TOTAL        80     5    15    100
               80.0   5.0  15.0  100.0
```

Kappa Statistics

```
Statistic          Value     StdErr    95% Conf. Limits
-------------------------------------------------------
Kappa              0.6765    0.0877    (0.5046, 0.8484)
Weighted Kappa     0.7222    0.0843    (0.557, 0.8875)
```

The Kappa and Weighted Kappa results are displayed, along with 95% confidence limits. Kappa generally ranges in value from 0 to 1 with a value of 1 meaning perfect agreement. (Negative values are possible.) The higher the value of Kappa, the better the strength of agreement.

Advanced ANOVA

Advanced ANOVA designs in WINKS include

- Two-Way ANOVA
- Two-Way with Repeated Measures
- Analysis of Covariance
- Three-Way ANOVA

Two-Way ANOVA
(Analyze/Advanced ANOVA/Two-Way ANOVA)

In a two way analysis of variance, the experimental design consists of two grouping factors and one or more observations on each combination of the grouping factors. For example, suppose you've designed an experiment to examine the effectiveness of several display strategies on sales. You have selected three display widths, and two heights, giving you 6 display combinations. In order to make comparisons of sales for each combination of height and width, you want to place one of the 6 display combinations at each of several stores. Then, after a period of time, you will examine the sales from each combination to see if you can discover which combination produces the most sales.

You decide to use each of the display combinations 4 times, which means you must have a total of 24 display locations. In order to prevent possible bias in choosing which combination to use where, you randomly choose where to place each combination, making sure that each of the 6 combinations is used 4 times. After a selected time period, you collect the data, which is number of sales per display. Your data is summarized as follows:

```
                       Display Width

Display
Height            Short    Regular    Wide
        Bottom     24        31        35
                   25        28        32
                   30        33        31
                   28        35        38

          Top      31        36        41
                   32        32        36
                   33        33        34
                   36        41        32
```

Create a database that contains 24 records, one record for each observation and three variables, HEIGHT, WIDTH, and SALES. Height is designated as B and T, and Width is designated as 1, 2 and 3. The data in the SALES.DBF database look like this:

HEIGHT	WIDTH	SALES
B	1	24
B	1	25
B	1	30
B	1	28
etc...		
T	1	33
etc...		
T	3	36
T	3	34
T	3	32

Step 1: Open the database named SALES.

Step 2: From the Analyze menu, select "Advanced ANOVA," then choose the Two-Way ANOVA option

Step 3: Choose HEIGHT and WIDTH as Group factors and SALES as the data value.

Step 4: You will be prompted to indicate if the grouping factors are FIXED or RANDOM. A FIXED factor is one that you have purposely chosen from a set of possibilities - such as the height and width for your displays. A RANDOM factor would be, for example, a factor where you chose the height and width randomly from a list of all possible heights and widths. The factors used in this experiment are fixed, since we specifically chose the heights and widths to test rather than randomly selecting them from a population of all possible heights and widths.

Step 5: A summary of the group means will be displayed as well as an ANOVA table.

Following the summary is the analysis of variance table:

```
Analysis of Variance Table

Source            S.S.   DF     MS     F   Appx P
---------------------------------------------------
Total           407.96   23
  Cells         220.71    5
  WIDTH         108.33    2   54.17  5.21  0.016
  HEIGHT         92.04    1   92.04  8.85  0.008
  INTERACTION    20.33    2   10.17  0.98  0.395
Within Cells    187.25   18   10.40
---------------------------------------------------
```

If the interaction effect is considered non-significant, multiple
comparisons of marginal means is appropriate. The first number
you should look at is the INTERACTION p-value. In this case, the
interaction effect can be considered as statistically non-significant
since the p-value for this test is 0.395. See the section titled
"Interaction Plots" for more information on how to interpret the
interaction effect."

When the interaction effect is non-significant, then it is appropriate
to compare the means of the main factors (WIDTH and HEIGHT)
directory (main effects). If the interaction effect is significant, then
it is appropriate to compare individual means (within the 6
combinations) rather than to compare HEIGHT for all WIDTHS
and WIDTHS for all HEIGHTS.

Continuing with this current example, to examine main effects,
you look at the p-value for WIDTH and HEIGHT. In this case,
both HEIGHT and WIDTH effects are significant at the 0.05 level
(since they both report p-values less than 0.05). This means that
both HEIGHT and WIDTH were important (at least statistically)
factors affecting the number of sales in a display. Your next
concern should be, "Which HEIGHT and which WIDTH produced
the most sales?"

Looking at the number of sales for HEIGHT, you can state that
sales from displays using the TOP location are statistically higher
than sales using the BOTTOM location.

Since there are more than two levels of WIDTH, you must perform multiple comparison tests to determine where the statistical significances lie. The program will give you an opportunity to do these comparisons. You will choose to compare the marginal means for WIDTH (3 means). Some of the results of the Multiple comparison test are as follows:

```
                                       Critical Q
Comp             Difference   P   Q      (0.05)
-----------------------------------------------
Mean(3)-Mean(1)  = 5.0       3   4.385   3.609*
Mean(3)-Mean(2)  = 1.25      2   1.096   2.971
Mean(2)-Mean(1)  = 3.75      2   3.289   2.971*
```

Comparisons marked with an asterisk "*" are significantly different at the 0.05 significance level (alpha-level). A graphic description of the multiple comparisons is given by:

Homogeneous Populations, groups ranked

```
        Gp Gp Gp
         1  2  3
            -------

    ----
```

```
This is a graphical summary of the Newman-Keuls
multiple comparisons test. At the 0.05 significance
level, the Means of any two groups underscored by the
same line are not significantly different.
```

In this case, you can conclude that the WIDTHS 2 and 3 are both better (more sales) than WIDTH 1, but no significant difference was found between WIDTHS 2 and 3.

Thus, your overall conclusion is that display sales are better in general at the top level, and better in general for widths 2 and 3 (regular and wide). However, there was not enough evidence to conclude that the wide width produced more sales than the regular width. Note: If you use Tukey or Scheffe, comparisons, your results may differ.

Interaction Plots

From the viewer, you can click on the Graph button to display an interaction plot. It is valuable in analyzing the experiment to look at the interaction of mean values across combinations of the factors.

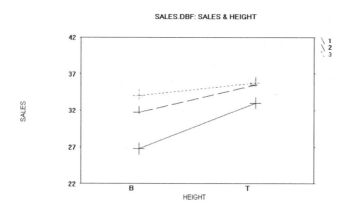

If the plots intersect, it usually means that an interaction effect exists. That is, the means behaved differently across levels of the factors. If the plots are fairly parallel, it means that no interaction effect exists. That is, the means behaved similarly across levels of the factor. If you examine the interaction plots for this example, you will see that the plots produce "almost" parallel lines -- at least they do not intersect.

Two-Way Unbalanced Design

An unbalanced design occurs when sample sizes for cells of the two-way ANOVA are not equal. WINKS uses a technique called the "regression approach" to perform calculations. When the data are balanced, you will get the same answers using this option as the balanced options. For more information about the unbalanced case, refer to the following references: Neter, Wasserman and Kutner (1990) , Kutner (1974) and Elliott and Woodward (1986). For those familiar with other statistical packages, this technique is the same as SPSS "Option 9" and SAS Type III SS.

Note: Care must be taken when you have unequal cell sizes. If the inequality across cells is great or if the inequality of cell sizes is due to some factor other than randomness, there may be serious violations of the underlying model.

Two Way Repeated Measure ANOVA
(Analyze/Advanced ANOVA/Tw0-Way Repeated Measures)

In a two-way analysis of variance, it is common to examine one "subject" at several points in time, or under several conditions. This differs from the replicates in the first example, which were all different locations. In the first example, if all four replicated for each combination of displays was in the same store, say observed on different weeks, then it would have been a "repeated measures design." Thus, the main difference between the first example and this example is that the "replicates" in the first example were unrelated, and in the repeated measures example, the replicates are related.

This example for a two-factor analysis of variance with repeated measures on one factor is taken from Winer, page 525 (see reference list). In this example there are two methods of calibrating DIALS (factor A), and the levels of B are four SHAPES of the dials. Six subjects were randomly assigned to perform the calibrating on a particular dial (A) for all four shapes of dials. That is, each of the six subjects were observed four times, once for each combination of the DIAL/SHAPE settings. The scores observed are accuracy.

The data is as follows:

Repeated Measures Data from Winer, Page 525

	SUB-	--SHAPES--			
A (DIALS)	JECT	B1	B2	B3	B4
1	1	0	0	5	3
1	2	3	1	5	4
1	3	4	3	6	2
2	4	4	2	7	8
2	5	5	4	6	6
2	6	7	5	8	9

For entry into a database, subjects are re-numbered as 1,2,3 and 1,2,3 . In this repeated measures design, the program will assume that subjects 1,2 and 3 in group 1 (A) are different than subjects 1,2, and 3 in group 2 (A). We are not necessarily interested in testing a subject effect in this experiment. Our objective is to determine if there is a Dial and/or Shape effect. Thus, the data in the database which will be called REPEAT2 will look like this:

GROUP	VARS	Repeated measures			
A	SUB	B1	B2	B3	B4
1	1	0	0	5	3
1	2	3	1	5	4
1	3	4	3	6	2
2	1	4	2	7	8
2	2	5	4	6	6
2	3	7	5	8	9

Step 1: Open the database named REPEAT2.DBF.

Step 2: From the Analyze menu select "Advanced ANOVA" then "Two-Way Repeated Measures ANOVA."

Step 3: Select variables A and SUB as your grouping factors, and variables B1, B2, B3 and B4 as the repeated measures.

NOTE: The order that you choose these in the program is important. Choose the variable on which the repeated measures are taken (in this case subject) as the second (2nd) in the list of grouping variables.

Step 4: A summary of the calculation results include this information:

```
Analysis of Variance
-------------------------------------------
Source                 DF    M.S.    F       P
-------------------------------------------
Between subjects       5
A                      1    51.04   11.89   0.03
Subjects in groups     4     4.29
Within Subjects        18
Repeated Measure       3    15.82   12.80   0.00
Interaction            3     2.49    2.01   0.17
Rep.Mes.xSub in gp    12     1.24
-------------------------------------------
```

If the interaction effect is considered non-significant, multiple comparisons of marginal means is appropriate. As in the previous example, look at the interaction effect first.

In this case, the interaction effect is not statistically significant ($p = 0.17$). This means that you may examine the main effects (A-Dials and Repeated Measures (B1, B2, B3, B4 - Shapes) test directly. In this case, both the dial effect and shape effect are statistically significant (at the 0.05 level).

Since there are 2 dials, you can immediately conclude that there is a difference in mean accuracy scores between dials. Scores for dial A1 are significantly lower than scores for dial A2.

To examine where significances lie in the repeated measures (shapes), you must perform multiple comparisons.

Comparisons in the multiple comparisons table marked with an asterisk "*" are significantly different at the 0.05 significance level (alpha-level). A graphic description of the multiple comparisons is given by:

Homogeneous Populations, groups ranked

```
        Gp     Gp    Gp    Gp
        B2     B1    B4    B3
                     ----------
        ---------
```

The conclusion in this case is that the mean scores for shapes 3 and 4 are higher than the mean scores for shapes 1 and 2. Furthermore, there are no statistically significant differences between shapes 1 and 2 and between 3 and 4. Note: If you use Tukey or Scheffe, comparisons, your results may differ.

If an interaction effect had been present, you would need to compare cell means rather than marginal means. For example, you would compare dial 1, shape 1 with dial 1 shape 2, etc. A multiple comparison test is provided to do these comparisons.

Interaction Plots

From the viewer, you can click on the Graph button to display an interaction plot. If the line on the plot intersect, it usually means that an interaction effect exists. If the lines are fairly parallel, it means that no interaction effect exists.

Three-Way ANOVA
(Analyze/Advanced ANOVA/ Three-Way ANOVA)

The Three-Way Factorial design has three grouping factors (independent variables) and one observed value (dependent variable). WINKS allows up to 5 levels of each of the grouping factors. The model for the analysis can be stated as:

OBS = A B C A*B A*C B*C A*B*C

where A, B, and C are main effects of the three factors. A*C, A*C

and B*C are the two way interactions and A*B*C is the three way interaction. OBS is the observed (dependent) variable.

The Analysis of Variance table reports the sum of squares and resulting F-test for each of the components of the model. Type III sums of squares are calculated, allowing unequal cell sizes. Empty cells are not allowed. Interpretive problems may arise from an analysis having unequal cell sizes. You should reference a good book to determine if this effects your analysis (Such as Neter, Wasserman and Kutner).

To interpret a three factor, first look at the three way interaction.

If it is not significant, then look at the two way interaction. If these are not significant, then you can examine the main effects tests. Differences between groups in main effects of over two levels can be analyzed using multiple comparison procedures. If three way interaction is present, analysis of the two way interaction terms or the main effects is invalid. If there are significant two way interactions, then tests for main effects contained in those interactions are invalid. In these cases, you must perform comparisons of means by cells, or remodel your analysis.

For example, use the file called 3WAYAOV.DBF to perform an analysis. The three factors variable as A, B and C. The A factor has four levels, B has 3 and C has 2. The observed variable is OBS.

Partial output for this analysis is given below:

```
Dependent Variable:OBS


Source          DF    Sum Sq     Mean Square    F      Value p
-------------------------------------------------------------------
Betw. Trt.      15    802.3788    53.49192     3.98     .001
-------------------------------------------------------------------
A                3    258.6133    86.20444     6.41     .002
B                1    367.1633   367.1633     27.31     .000
C                1     87.65879   87.65879     6.52     .016
A*B              3     50.06533   16.68844     1.24     .313
A*C              3      7.615718   2.538573     .19     .903
B*C              1     68.70384   68.70384     5.11     .032
A*B*C            3     54.02666   18.00889     1.34     .281
-------------------------------------------------------------------
ERROR           28    376.4167    13.44345
-------------------------------------------------------------------
TOTAL           43   1178.795
```

Three-way analysis output also includes summary statistics by group. If needed, you can use the Multiple Comparison analysis to perform comparisons of means.

Advanced Regression Procedures

Five WINKS Advanced Regression options helpful in select the model regression appropriate for a data set are:

- Polynomial Regression
- All Possible Regressions
- Stepwise Multiple Linear Regression
- Logistic Regression
- Bland-Altman Analysis

Polynomial Regression
(Analyze/Advanced Regression/Polynomial Regression)

Polynomial regression is used when the relationship between predictor and response variables is curvilinear. The data given in the following table are the ages of 29 players and their scores on a new video game (generated data). The relationship between AGE and SCORE is not linear, but that a quadratic term may be helpful in describing the relationship. That is, a model such as

$$Y = b_0 + b_1X + b_2X^2 + e$$

might be considered in this case. The GAME.DBF database
(partial listing):

RECORD	AGE	SCORE	CENTERED
1	6.9	24710	-11.75
2	7.4	26730	-11.25
3	7.9	25920	-10.75
4	8.3	27510	-10.35
:	:	:	:
28	41.6	9850	22.95
29	43.1	6000	24.45

Such a polynomial model can be recognized as a form of a
multiple linear regression model with two predictor variables, X
and X2.

In fitting a polynomial regression model, all lower order terms
must be included. That is, the first-order term is used, and higher
order terms are used only if the first-order term is not sufficient. A
cubic term is used only if both the linear (first-order) and quadratic
(second-order) terms are included. When using Polynomial
Regression in WINKS, you are asked to specify the order of the
polynomial you wish to fit.

As with any multiple regression analysis, care must be taken to
avoid collinearities between the predictors. That is, if the
predictors are highly correlated, the coefficient estimates may
contain considerable error. (See, e.g., Montgomery and Peck,
1982, p.184) Since X and X2 may be highly correlated the
collinearities can be reduced by expressing these predictors as
deviates from the sample mean. That is, define the predictors as
$(X-Xbar)$ and $(X-Xbar)^2$.

Centering the data may not always sufficiently reduce
collinearities, in which case the data should be standardized (divide

the centered values by the standard deviation of the predictor variable values).

There are various approaches for determining the order of the model. One method is a "forward selection" procedure in which the first-order (linear) term is fit and then higher order terms are added sequentially until the F-test for a non-zero coefficient is not significant for the highest order term. Another method is a "backward elimination" procedure in which an appropriately high-order polynomial model is fit and terms are deleted one at a time from high to low order until the highest order term of the remaining terms results in a significant F-test. These two methods may not result in the same model.

WINKS fits a model of the order you select and reports the coefficients of each term, including an intercept term, up to that order. The results of the tests of significance of these coefficients are also reported. A small p-value indicates that the corresponding coefficient is significantly different from zero. Residual analysis is also useful for investigating the appropriateness of the model selected. WINKS also reports the Analysis of Variance for the entire regression fit, as well as R-Square and adjusted R-Square, as it does in the regular Linear Regression module.

In general, in regression analysis simpler models are preferred. It may be possible to transform the predictor in some way so that higher order terms are not necessary. Terms higher than second or third order are not usually used unless there is some reason. *It is always possible to fit a high enough order model, but such a model is difficult to interpret and not generally recommended.*

As with any regression model, extrapolation is risky and should be avoided. While a polynomial model may adequately model the relationship between variables within the range of the data used in the analysis, it is extremely risky to assume that relationship continues to exist outside the range of the data. Refer to a standard text, such as Neter and Wasserman or Montgomery and Peck, for more information about polynomial regression.

As noted earlier, in order to reduce collinearities, the predictor variable values are standardized or, in this case, centered. That is, the mean of AGE, 18.65, is subtracted from each value of AGE to create a new variable, CENTERED, also contained in GAME.DBF. While the correlation coefficient of AGE and AGE2 is 0.98, that of CENTERED and CENTERED2 is 0.78, so CENTERED is used in the regression analysis. Follow these steps to perform this analysis:

Step 1: Open the database named GAME.DBF.

Step 2: From the Analyze menu, select "Advanced Regression, " then "Polynomial Regression Analysis."

Step 3: Select the field SCORE as the dependent variable and select CENTERED as the independent variable.

Step 4: Select "2nd order quadratic regression." Results are as follows: The first-order term, CENTERED^1, has an F-statistic of 11.51 (p=0.002) and the second-order term has an F-statistic of 292.58 (p<0.001). The quadratic term is highly significant; it adds greatly to the model after the first-order term has been fit.

Since the quadratic term is significant, try adding a cubic term using "3rd order, cubic regression." The results of a cubic regression show the F-statistic for CENTERED^3 to be 1.65 (p=0.210), is not significant, and hence the cubic term is not useful to the model. Thus the selected model is

SCORE = 28645.88 - 91.68(AGE-Mean AGE) - 32.67(AGE-Mean AGE)^2 .

To express the model as SCORE = b0 + b1(AGE) + b2(AGE)^2 , the relationships between the two sets of coefficients

$$b_0 = b_0{}^* - b_1{}^*X + b_2{}^*X^2$$
$$b_1 = b_1{}^* - 2b_2{}^*X$$
$$b_2 = b_2$$

are used, where $b_0{}^{*\prime}$, $b_1{}^*$ and $b_2{}^*$ are the coefficients in the model using the CENTERED predictor variable. In this example, then,

b0 = 28645.88 + 91.68(18.65) - 32.67(18.65)2 = 18992.35
b1 = -91.68 -2(-32.67)(18.65) = 1126.91
b2 = -32.67

and the model in terms of AGE is

SCORE = $18992.35 + 1126.91(AGE) - 32.67(AGE)^2$.

Using this regression equation to predict the response, SCORE, given values of the predictor variable, AGE, the score for someone 32 years old is predicted to be 21599.39 while that of a 16-year-old is 28659.39. To display a plot of the regression, click on the GRAPH button and select to plot "Centered by Score." The graph is down below:

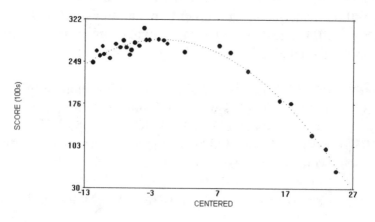

Stepwise Selection for Multiple Regression
(Analyze/Advanced Regression/Stepwise Regression)

For a large number of predictors, or if for other reasons the All Possible Regressions variable selection procedure is not practical (see All Subsets Regression description below), an alternative is the Stepwise variable selection procedure. WINKS Stepwise option can consider up to 49 variables, and can define a model using up to 20 of those variables. As noted earlier, the Stepwise procedure is a combination of "forward selection" and "backward elimination" techniques.

At the first step, the model consisting of all variables is considered, and the variable testing "most significant", i.e., having the largest F-statistic, becomes the first variable included in the model. In the second step, the variable selected in the first step is forced into the model and the other variables are then fit. A cut-off p-value is used as the selection criteria to determine whether any more variables should be included. This cut-off p-value selection criteria can be designated by you, or else the default criteria used by WINKS is a p-value of 0.25 for the F-tests. Of those variables meeting the selection criteria at step two, the one showing the most significance, i.e., having the largest F-statistic, is added to the model consisting of the variable selected in the first step.

The two-variable model is then "checked" and if the coefficients of both variables are shown to be significantly different from zero (having small p-values), the process continues. Again, the cut-off p-value can be set by you, or else the default is 0.25. At the third step, the two already chosen variables are forced into the model and the other variables then fit. If any remaining variables meet the selection criteria, the "most significant" of those is added, and the three-variable model checked. The process continues as long as all selected variables satisfy the "checking" procedure and as long as at least one remaining variable meets the selection criteria and is added to the model at each "forward" step. The operator is also given the opportunity at each step to continue or to stop the procedure.

The data used in this example are contained in the CRIME.DBF database. Each of the 141 records contains U.S. Census Bureau information on one metropolitan area in one year. The response variable, CRIMES, is the total number of crimes. There are nine predictors:

```
AREA: number of square miles
POP: total population
CITY: percent of population in central cities
OVER65: percent of population age 65 and older
DRS: number of active physicians
HOSP: number of hospital beds
HSGRAD: percent of adult population having completed
high school
LABOR: number of persons in civilian labor force
INCOME: total income received
```

Preliminaries (Transformations, Indicator Variables, Outliers)

Scatterplots of the quantitative predictors against the response (which can be easily displayed using the graphical correlation matrix procedure) raise doubts about the linearity of some of these relationships. Natural log transformations of the quantitative predictor variables as well as the response variable result in approximately linear relationships. Therefore, these transformed variables are used in the analysis, and are included in the database, CRIME.DBF, as LNCRIMES, LNAREA, LNPOP, etc

In this example, none of the observations has been excluded from the analysis, but there are a few data points which might be considered questionable. There are a few observations with exceptionally large values of the response variable. It is a difficult judgment whether to exclude observations from the analysis and such action should be taken only with justification.

Refer to a standard regression textbook, or especially to Belsley, Kuh and Welch (1980), for discussion of techniques for identifying influential observations, or outliers, and for discussion of other considerations which are preliminary steps to variable selection. For example, follow these steps to perform a Stepwise analysis:

Step 1: Open the database LNCRIME.DBF. It contains log values of the fields to be used from the original CRIME.DBF database.

Step 2: From the ANALYZE menu select Advanced Regression then "Stepwise Regression".

Step 3: You will be prompted to enter which fields to use. Select LNCRIMES as the dependent variable, and LNAREA, LNPOP, LNCITY, LNOVER65, LNDRS, LNHOSP, LNHSGRAD, LNLABOR and LNINCOME as nine independent variables.

Step 4: You are prompted to indicate any variables you want to force the model to include. Simply press Enter to indicate none. Then you are asked to specify the cut-off p-values for adding variables in the "forward" steps and for dropping variables in the "backward" steps. Press enter to select the defaults of 0.25 in both cases.

Step 5: WINKS begins by performing the regression using the full set of nine predictors, and selects LNPOP as the "most significant" variable (F=598.86, p<001). Continuing the procedure, WINKS considers the eight tow-variable models, each consisting of LNPOP and one of the other eight predictors. Of these, the variable LNHSGRAD has the largest F-statistic (11.61, p=.001) when fit after LNPOP, so LNHSGRAD is added to the model previously consisting only of LNPOP and the constant term.

If you continue the procedure, WINKS then tests this two-variable model to make sure that the term previously included, LNPOP, remains significant after LNHSGRAD is added. the coefficient of LNPOP now has an F-statistic of about 691.82, p, so neither variable is eliminated in the first "backward" step. Continuing, the next forward step considers the eight three-variable models, each consisting of LNPOP, LNHSGRAD and one of the other remaining seven predictors. LNLABOR is found to have the largest F-statistic (2.83, p = 0.099) and is added to the model.

Again, the backward procedure tests this model consisting of LNPOP, LNHSGRAD and LNLABOR, but does not eliminate any of them since none of the corresponding F-tests result in p-values greater than 0.25.

At each step you are asked to continue or stop the procedure. There is some concern in this data about collinearities between some of the predictors. It would be advisable to display pairwise correlations of the variables using the WINKS Regression module. If two highly correlated variables enter the equation, you might want to run the stepwise procedure again, leaving one of them out. The final model selected by this run of the Stepwise procedure is:

```
LNCRIMES = .0150363 + 1.5690325 (LNPOP) + 0.7512479
(LNHSGRAD) - .4524823 (LNLABOR)
```

with R-Square equal to 0.9426, adjusted R-Square equal to 0.9389, and MSE equal to 0.038.

All Subsets Regressions (Analyze/Advanced Regression/All Possible Subsets)

Also known as "best subset selection", this procedure consists of considering all possible combinations of the predictor variables. It is then possible to compare all possible models and choose the "best" one. Comparison can be based on a number of criteria, including mean squared error, Mallow's Cp, and R-Square.

The calculated MSE is an estimate of the variance of the errors in the full model. A smaller error variance is desirable, so different models can be compared based on MSE, with those having a smaller MSE preferred.

R-Square, the coefficient of determination, is a measure of how much of the variability in the response is explained in the model, provided the model has been arrived at properly. A model with larger R^2 is preferred to one with a much smaller R^2.

Mallow's C_p is a statistic which is a function of the error sum of squares for the full model and that for the reduced model. The formula for C_p is $(SSE_p/s^2)-(N-2p)$, where SSE_p is the error sum of squares for the reduced model with p terms, s^2 is the estimate of MSE for the full model and N is the number of observations. Under the correct model, C_p is approximately equal to p and otherwise is greater than p, reflecting bias in the parameter estimates in the regression equation. Thus, it is desirable to select a model in which the value of Cp is close to the number of terms, including the constant term, in the model.

These three criteria are typically used to compare combinations, or subsets, of the predictor variables. When WINKS reports the results of the All Possible Regressions procedure, it reports all three of these criteria. Of course, you should also take into account any theoretical criteria specific to the problem for including or excluding variables, as well as be careful not to include redundant variables, which may introduce collinearities. It is often helpful to consider which variables consistently appear in the better models. The better models can then be analyzed using the Multiple Regression option of Regression and Correlations, and the results of tests for significant coefficients considered in the final decision. Residual plots of predicted values under the chosen model should show a random scatter of points.

Clearly, comparing all possible models is generally the best method for making a decision about a "best" model since it provides the most information about the available choices. However, the "all possible" subsets procedure can become quite

large with just a moderate number of predictors. WINKS has the
capability to perform All Possible Regressions on a maximum of
eight predictor variables. With eight variables, there are 2^8-1, or
255, possible subsets, and the procedure can take some time.

As an example of the All Possible Regressions procedure, consider
the Longley data. Follow these steps to perform this analysis:

Step 1: Open the database named LONGLEY.

Step 2: From the ANALYZE menu, choose Advanced Regression
then "All Possible Regressions".

Step 3: When asked to specify the fields to use, select TOTAL as
the dependent field and all others as independent fields.

Step 4: WINKS reports the results of all six single-variable
models, then all 15 two-variable models (15= six taken two at a
time), 20 three-variable models, and so forth. For each of the 26-
1=63 models, WINKS reports p, the number of variables in the
model including the intercept term, the degrees of freedom, sum of
squares for error and mean-squared error for the full model, R-
square and Mallow's Cp.

A partial listing of the eight models with lowest MSE, highest R-
square and Cp approximately equal to p, are listed below. All of
the 55 other models have very large values of Cp.

Model	R^2	Cp
(UNEMP, ARMED, TIME)	.993	6.2
(DEFLATOR, UNEMP, ARMED, TIME)	.993	8.2
(GNP, UNEMP, ARMED, TIME)	.995	3.2
(UNEMP, ARMED, POP, TIME)	.995	4.6
(DEF., GNP, UNEMP, ARMED, T	.995	5.1
(DEF., UNEMP, ARMED, POP, TIME)	.995	6.1
(GNP, UNEMP, ARMED, POP, TIME)	.995	5.0
(DEF., GNP, UNEMP, ARM, POP, TIME)	.995	7.0

Of these eight models, all include variables #3,4,5 (UNEMP,
ARMED, TIME). The pairwise scatterplots and pairwise

correlation coefficients of the six predictors show that TIME, DEFLATOR, GNP and POP are all highly correlated (r>0.90). Therefore, only one of these four variables needs to be included in the model. Furthermore, note that the models with more variables than UNEMP, ARMED and TIME have only trivially larger values of R2. It is helpful to display the correlation matrix for these variables to see how variables are correlated. The multiple regression on the model including UNEMP, ARMED and TIME confirms all three coefficients to be significant. Therefore, an appropriate model is:

```
TOTAL = -1797221.11 - 1.47(UNEMP) - 0.77(ARMED) +
956.38(TIME)
```

Use the multiple regression procedures to run a full regression analysis on these terms.

Simple Logistic Regression
(Analyze/Advanced Regression/Logistic Regression)

Logistic Regression is used to analyze the relationship between two variables when the dependent variable is binary. This differs from normal simple linear regression where the dependent variable is a continuous numeric variable. The logistic regression model can be described by

$$\text{logit}(p_i) = \log(p_i / (1 - p_i)) = \beta_0 + \beta_1 * x_i$$

where

p_i is the response to be modeled
β_0 is the intercept parameter
β_1 is the slope parameter
x_i is an array of independent variables

The logistic model uses the logit transformation of the ith observation's event probability, p_i , as a linear function explained by the independent variables x_i. Thus, for a binary dependent variable and a continuous independent variable, the WINKS program will calculate the coefficients for the logistic equation that best fits the data.

In WINKS, there are two ways to enter data for use in the logistic procedure. You may enter your data as raw data or summarized data. In the summarized data method, you need at least three fields in your database — The independent variable (X), the number of observations for each value of the independent variable (N_j), and a count of positive outcomes from the dependent variable. For example, suppose you are testing coupons that offer discounts of 5, 10, 15, and 20 percent off. You give away 400 of each kind of coupon and observe how many are redeemed.

Xj = discount value of coupon
Nj = 400 for each value of the coupon
Cj = How many coupons for value j were redeemed

The program will calculate the proportion of coupons redeemed (Pj) for the information above. For example, suppose your data for this experiment is as follows:

Discount	Given out	Redeemed
5	400	57
10	400	93
15	400	145
20	400	209
30	400	305

To analyze this data, follow these steps:

Step 1: Open the database named logistic.dbf. This database contains the data in the table above.

Step 2: From the *Analyze* menu, select Advanced Regression, then Logistic.

Step 3: You must carefully select the field names in the correct order. First select DISCOUNT, then click Add.

Step 4: Select GIVEN, then click Add.

Step 5: Select USED, then click Dep. Var.

Step 6: Click OK. Optionally enter numbers for use in prediction. You will be given a chance to enter values you want to predict, and the calculations will be performed and reported in the output. (See discussion below.)

```
Dependent variable is USED
Independent variable is DISCOUNT
Weights variable is GIVEN
Number of cases is   5
```

Variable	Coefficient	St. Error	t-value	p(2 tail)
Intercept	-2.361179	.0587289	-40.20472	<.001
GIVEN	.1193188	.0031445	37.945333	<.001

The fitted transformed logistic response function is

P' = -2.361179 + .1193188 * DISCOUNT

Step 7: Using the following equation, estimate the percent of redemption from a 15% coupon:

$$P = \frac{e^{\beta_0 + \beta_1 * \text{Discount}}}{1 + e^{\beta_0 + \beta_1 * \text{Discount}}}$$

where

$$\beta_0 + \beta_1 + \text{Discount} = -2.361179 + .1193188 * 15$$
$$= -0.571397$$

Putting the –0.571397 into the equation yields the value

$$P = 0.360915$$

Thus, you estimate that about 36% of the 15% off discount coupons will be redeemed.

Using Raw Data In Logistic Regression

A second way to read in data in the Logistic Regression procedure is to read in only two fields — the independent variable (X) and a 0/1 (binary) dependent variable. For the coupon data, this database would look something like this:

COUPON	USED
5	1
5	0
15	0
15	0
Etc...	1

In this database, each coupon has an entry, so for each of the 5, 10, 15, 20 and 30 percent off coupons, you have one record, making a total of (400*5) = 2,000 records. If your data is in this raw form, use the Tabulation procedure to calculate counts for each group.

For example, using the data in lograw.dbf file, you will get the following table:

```
-----------------------------
|          |      USED        |
|          |------------------|
|          |   0   |    1     |
|--------|---------|----------|
|COUPON  |         |          |
|--------|         |          |
|5       |    343|        57|
|--------+---------+----------+
|10      |    307|        93|
|--------+---------+----------+
|15      |    255|       145|
|--------+---------+----------+
|20      |    191|       209|
|--------+---------+----------+
|30      |     95|       305|
|--------+---------+----------+
```

Use this information in column 1 to create a database usable for the logistic procedure as shown in the preceding example.

Bland-Altman Plots

(Analyze/Advanced Regression and Comparison/Bland-Altman)

A Bland-Altman analysis is a way to assess agreement between two methods of clinical measurement. Using an example from a paper by Bland and Altman (1986), suppose there is a measurement (or peak expiratory flow rate/PEFR) made from using "Large" meter and a "Mini" meter. You want to know if they measure the same thing.

Two plots are used to analyze this data, a plot of identity and a Bland-Altman plot. The data follows:

SUBJECT	LARGE	MINI
1	494.0	512.0
2	395.0	430.0
3	516.0	520.0
4	434.0	428.0
5	476.0	500.0
6	557.0	600.0
7	413.0	364.0

8	442.0	380.0
9	650.0	658.0
10	433.0	445.0
11	417.0	432.0
12	656.0	626.0
13	267.0	260.0
14	478.0	477.0
15	178.0	259.0
16	423.0	350.0
17	427.0	451.0

The first is called a plot of identity. It is similar to a regression plot, except the line in the plot is a line based on X = Y. If the scatter of points in this plot lies near the line, it indicates that the two ways of measuring PEFR are similar. To display this plot:

Step 1: Open BAEXAMP.DBF. Select Analyze/Advanced Regression and Comparisons/Bland Altman/Identity Plot.

Step 2: Select the two variables LARGE and MINI. The following plot is displayed:

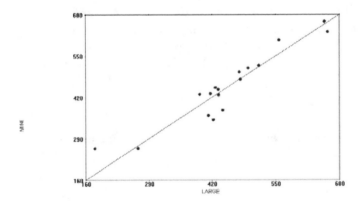

From this plot of identity, you can see that the scatter of points falls close to the line.

A second method of looking at this data is with a Bland-Altman Plot. This plots the average of the two values by subject against the difference. 95% confidence limits are also calculated and displayed on the graph (Mean difference ± 2 * Standard Deviation)

To display this plot (after opening the BAEXAMPLE DATA):

Step 1: Select Analyze/Advanced Regression and Comparisons/Bland Altman/Bland-Altman Plot.

Step 2: Select the two variables LARGE and MINI. The following plot is displayed:

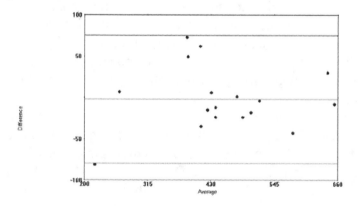

If the two measures are measuring the same thing in a unbiased way, this plot should show a random scatter of points between the upper and lower confidence limits. The plot above is indicative of a good fit. If there is a pattern to the points, there might be some biased associated with the measurements.

Time Series Analysis
(Analyze/Time Series Analysis)

The WINKS Times Series module allows you to perform model identification, estimation and forecasting. Data are sometimes observed as a series of numbers over a period of time. Some common series of data include sunspots, airline passengers per month, monthly sales figures, stock prices and the like.

Description of the Time Series Process

Time series analysis deals with attempting to model an observed series of data points to forecast future activity or to understand the driving mechanism. There are a number of approaches to modeling. This time series program bases its modeling techniques on the ARMA (autoregressive moving average) approach. In this approach, the researcher must first decide if there is an autoregressive (AR) and/or moving average (MA) component, and the order of each. These orders will be called p and q. We will use p as the order of the AR component and q as the order of the MA component. Thus, a model will be designated as an ARMA(p,q). For example, the model ARMA(8,0) means that the order of the AR component is 8 and the order of the MA component is 0 (none). The goal is to find a model which adequately describes the process without using any unnecessary parameters, a parsimonious model.

For those with a mathematical bent, the ARMA model can be written mathematically as follows:

$$X_t = c\, X_{t-1} + \ldots + c\, X_{t-p} + a_t - d_1 a_{t-1} - \ldots - d_q a_{t-q}$$

where p and q are the model orders,

a_t is a zero-mean, white noise process, whose variance is called the white noise variance (WNV).

c_1 to c_p are the AR parameters
d_1 to d_q are the MA parameters
X_t is the data at time t

For this process to be stationary the roots of the characteristic equation $(1 - c_1 r - ... - c_p r^p = 0$ where r is a complex number) must lie outside the unit circle. See Box, Jenkins and Reinsel (1994) for a detailed explanation.

How To Analyze Time Series Data

The purpose of the WINKS Time Series program is to help you:

A) Decide what ARMA model is appropriate for your data
B) Estimate the parameters of the model
C) Create a forecast

A) Decide what ARMA model is appropriate for your data:
The first part of the analysis process is model identification. There are no easy answers to how your data should be modeled. Although much research is being conducted to help you identify a model, it still remains somewhat of an art. Here are a few items that you need to consider in choosing a model.

Is the data simply white noise? That is, do the values of the data go up and down in a completely random fashion? One way to determine if the data are white noise is to examine the sample autocorrelations. If they are small and uncorrelated then the process may be white noise. If the process is white noise, then only about 5% of the sample autocorrelations (absolute values) would be expected to be greater than 2 / sqrt(n) where n is the length of the series. Thus, for a series of length 100, you would begin to suspect that white noise is NOT a sufficient model if many more than 5 autocorrelations are greater than $(2/10) = 0.2$ in absolute value. If the sample autocorrelations are greater than the 5% limits for the first few lags, this indicates that modeling should be continued.

If the data are not white noise, you may then attempt to model it as an ARMA process. The WINKS program provides a statistic to help you decide what model is appropriate. The W-statistic (see Woodward and Gray) technique examines the data for fit to a series of models, and returns the three "best" guesses for a model. These are not necessarily the best models, but the technique can be helpful in choosing which models to examine.

There are a number of other techniques to assist you in choosing a model. Box and Jenkins recommend identification of p and q on the basis of examination of the sample autocorrelations and partial autocorrelations.

It should be noted that special care must be taken into account when some roots of the characteristic equation are near the nonstationary region. In this case the process may need to be transformed to a stationary model before the stationary components of the model can be seen. Please reference a standard time series text such as Box, Jenkins and Reinsel (1994) for further advice.

B) Estimate Parameters: Once a model has been chosen, you may estimate the values of the parameters of the model given your set of data. The WINKS program uses techniques developed by Tsay, Tiao and Burg to calculate the estimates of the model. (Tsay and Tiao, 1984)

 C) Forecast: Once you have the estimates of a model, you can use this information to create a forecast. During your process of deciding what model you will use, you may choose to forecast the last few known values of your series using the model under consideration. If the model estimates these to your satisfaction, then it may be a good model for forecasting into the future. WINKS will allow you to use your model to forecast future values of the series. An optional 95% confidence bound may be plotted for your estimated forecast.

The Modeling Process

The time series module allows you to perform some of the most common series evaluation techniques on a set of data, and will also allow you to estimate the parameters of an ARMA model (AutoRegressive Moving Average). Once a model is determined, the program allows you to create a forecast.

Step 1: Open the database named SERJ.

Step 2: From the Analyze menu select the "Time Series" option.

Step 3: Select Y as the variable to analyze. The Time Series Analysis window will be displayed containing Model, Estimate and Output menus.

Step 4: Examine the plots of the data, autocorrelations and partial autocorrelations. Display plots by selecting the Graph options from the Model menu in the Time Series Analysis Window. A plot of partial autocorrelations is shown below.

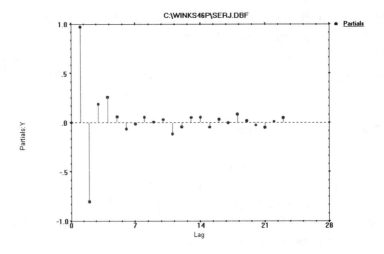

Step 5: Display tables of sample autorrelations and partial autocorrelations by selecting these options from the Model menu. (Refer to the tables below.) Notice that the sample autocorrelations are substantial, with the magnitude of many of them greater than the calculated 95% limit (described earlier in this section)

```
2 * ( 1 / 17.20) =   0.12
```

indicating that the data are not from a white noise process.

Autocorrelations:

Lag	Autocorrelations		
---	----------------	Std. Dev =	3.197
0	1.000	Mean =	53.509
1	0.971		
2	0.896		
3	0.793		
4	0.680		
5	0.574		
6	0.485		
7	0.416		
:			
etc			
:			

```
17                    0.140
18                    0.121
19                    0.110
```

Partial Autocorrelations:

```
Lag        Partial Autocorr
---        ----------------
 0             0.000
 1             0.971
 2            -0.804
 3             0.188
               :
              etc
               :
17            -0.003
18             0.085
19             0.017
```

Step 6: From the Model menu select "Model Identification/W-Statistic." In addition to the plots of the autocorrelation and partial autocorrelations, you can use the W-statistic to help you choose an appropriate model. That is, decide the values of P and Q for an ARMA model, where P represents the degree of the AR (auto-regressive) part of the model and Q represents the MA (moving average) part of the model. Give the program MAXIMUM values of P and Q to be considered. In this case, enter 10, 10. This means that you will consider an ARMA model up to P=10 and Q=10. The displayed results are:

```
Top 3 choices for P and Q using the W-statistic:
Choice  1) P=  3   Q = 2     W - statistic =  0.253
Choice  2) P= 10   Q = 8     W - statistic =  0.256
Choice  3) P=  3   Q = 9     W - statistic =  0.420
```

The lower the W statistic, the stronger the choice. Thus, from these results, the model 3,2 has the lowest value for W, and is therefore a strong choice. The W-Statistic technique will assist you in choosing a P and Q for an Autoregressive Moving Average ARMA(P,Q) time series model. This is only a tool in deciding P and Q, it does not necessarily give the 'best' model.

Step 7: From the Analysis menu, select the "Estimate Parameters" option to estimate model parameters. In this case use $p = 3$, $q = 2$. The results are:

```
Estimated AR Parameters:

  1 )    2.216831
  2 )   -1.770174
  3 )    .5256025

Estimated MA Parameters:

  1 )    .095701
  2 )   -.2602345
```

These parameter estimates may then be used to forecast new values.

Step 7: From the Analysis menu, select "Forecast Beyond Series." For this example, select 20 steps to forecast. The forecast values will be displayed. A partial list of the output is shown below. This table contains the lower and upper 95% confidence values as well as the forecast and the actual series.

COUNT	LOWER	UPPER	FORECAST	SERIES
1	53.8	53.8	53.8	53.8
2	53.6	53.6	53.6	53.6
	... etc ...			
296	57.0	57.0	57.0	
297	56.1393	57.4385	56.7889	
298	55.1388	58.1855	56.6622	
299	54.0181	58.896	56.4571	
	... etc ...			
305	50.4607	59.923	55.1919	
306	50.2187	59.7934	55.006	
307	49.9435	59.5997	54.7716	
308	49.5905	59.3138	54.4522	
309	49.3191	59.1023	54.2107	
310	49.038	58.8759	53.9569	
311	48.8671	58.7538	53.8105	
312	48.8726	58.8012	53.8369	
313	48.7651	58.7281	53.7466	
314	48.5294	58.5194	53.5244	
315	48.3593	58.3702	53.3648	
316	48.1499	58.1766	53.1633	

Step 8: From the Analysis menu, select "Display Graph of Forecast". A plot of the forecasts with 95% confidence limits will be displayed. See the plot below.

Note: You might try the other two recommendations from the W-Statistic to see if they produce a better forecast. Not all models suggested by the W-statistic will be well-behaved stationary models and thus these forecasts may be poor. As mentioned earlier, there is no guaranteed "best" solution for modeling a process. You must try several models until you find one that (hopefully) fits the data.

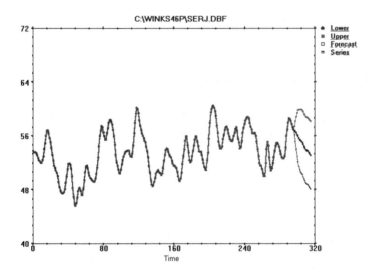

Note: The W-statistic will never select a strictly moving average model (i.e., p = 0.) For discussion of identification in this case see Box, Jenkins and Reinsel. Also, the W-statistic is appropriate for model identification for stationary models. When data are non-stationary or nearly non-stationary, differencing or other transformations to stationarity may be required.

Notes on Time Series Analysis

Other Time Series Analysis procedures or options that have not
been described above are discussed here:

Output Options

The Time Series Analysis procedures display both textual and
graphical output. Textual output is displayed on the screen, and is
automatically stored in the output buffer file. When you exit the
Time Series Analysis, all of the textual output that appeared during
the analysis will be displayed in the WINKS viewer -- the same as
output from any other analysis. Also, you can display the output at
any time by selecting the "View Text Buffer Contents" from the
Output menu.

When graphical output is displayed, you can choose to print, save,
or capture the image the same as in other WINKS graphical
displays. The options button on the graphical display allows you to
modify titles for the graph, and select other options.

Differencing the series

When a series is not stationary, you can sometimes produce a
stationary process by differencing the series. A new series is
created from the old by subtracting the observation d time units
from the original value. (i.e., you create Dt given by $Dt = Xt - Xt-d$)
For example, a difference using $d=1$ creates a new series that
consists of the difference between each value in the series minus
the previous value. This option allows you to choose the difference
order, d. You may also choose to save the differenced series in an
ASCII file that could later be imported into a database using the
WINKS ASCII import option.

After differencing, the new series (containing fewer points
according to the amount of differencing) can be used for analysis.

Note that differencing DOES NOT change the original data in the database. For more information on when to difference a series, refer to a text such as Box, Jenkins and Reinsel (1994).

Output Forecast Information to a File

You can use the "Output Forecast information to a file" option from the Estimate menu to save the forecast information to a comma delimited ASCII file, that could then be imported into a dBase file for further analysis using the WINKS ASCII import option.

Quality Control Charts

"You cannot inspect quality into a product." Harold F. Dodge (1893-1976).

Instead of being a final check to make sure a product has an acceptable number of flaws, Statistical Quality Control (SQC) is more correctly used as a tool to discover and correct current problems along the entire life of a product, from conception to customer support.

SQC control methods should be used to identify unexplained variation in a process. When such variation is identified, that variation can then often be controlled, corrected, or eliminated. The processes used in WINKS are a result of work of SQC pioneers including W. Edward Demming, Joseph M. Juran and Dr. Kaoru Ishikawa.

WINKS' QC Features

WINKS' QC tools can be a valuable part of making an overall QC plan work. Generally, the portion of the quality control system that is addressed by WINKS is that of estimating process variability. A process is examined by taking samples of the process over a period of time. Using certain criteria that produce limits of acceptability, a process is said to be in statistical control as long as the measured item stays within acceptable limits. If the item measured goes beyond acceptable limits, the process is said to be out of statistical control.

For example, in the production of a microprocessor chip, the company is interested in how many of the chips are acceptable, and how many fail an initial test. If the proportion of failed components grows beyond an acceptable limit, it may signal a failure of the process along the production line. Other tests along the line may be instituted to locate the source of the problem.

Another example might be a critical measurement on a bearing. The company buying the bearing demands that the bearing meet a certain tolerance--that is the size of the bearing may not vary by more than a very small amount. If the bearing is shipped, and the customer finds that some bearings are too small or too large, the entire shipment may be refused. Therefore, it is imperative that the manufacturer keep a close eye on the process. It would be too costly to measure every bearing, so a sampling scheme is devised, and a chart is produced that shows the average dimension of the sample bearings over time. If the chart reveals that some of bearings are beyond the statistical limits, manufacturing may be halted until the problem is found and corrected.

The WINKS Quality Control module allows you to create five basic kinds of control charts. They are the X-Bar chart, the R-Chart, the S-Chart, the P-Chart and the Individual Measurements Chart.

The X-Bar chart is a chart of the mean value of a sample taken over time. For example, in the case of the bearing manufacturer mentioned above, a sample of 5 bearings is taken from the manufacturing process once every 15 minutes during the day. The diameter of the bearings are immediately measured, and the average of the five bearing diameters are plotted on a chart. Statistical Control limits describe the limits of natural variation (i.e., based on within subgroup variation). If the average diameter of the bearings goes above or below the statistical limits, the manufacturing process is immediately halted, and engineers begin examining the machines to find the source of the problem.

Sometimes the average is not a sufficient measure for a process. For example, if the bearings are "on the average" okay, but the size varies widely, then this may give the manufacturer important information about the variability of the production process. Two charts provided to evaluate "within" sample variance over time are the R-Chart (Range Chart) and the S-Chart (Standard Deviation Chart). In the example above, along with X-Bar chart, an R-Chart or S-Chart may be produced. The Range chart shows the range (max-min) for each sample subgroup of 5 bearings. An S-Chart would show the variability of the sample based on the standard deviation of the 5 diameters. If the variability is large (even if the process is "in control" according to the X-Bar Chart), the management may want to examine the process and discover the cause of the excess variability. You might want to include some sort of sampling further down the production line to catch errors at their point of origin rather than at a later checkpoint. See "X-Bar and R-Chart Example" later in this topic.

Note: R-charts and S-charts are a graphical form of the test for "constant variance", which is a requirement for an Analysis of Variance test.

The P-Chart measures a proportion rather than a mean. It is intended for use on a process in which there are counts of "failure". For example, in the microprocessor example, a failure is the inability of a chip to pass a performance test. Since the manufacturer may be producing thousands of chips an hour, it would be very costly to have a large proportion of bad chips installed in a component, only to find that the component does not work because of the bad chip. A process of testing the chips can be devised so that the number of bad chips is kept to a minimum (hopefully near zero). If the proportion of failed chips rises above a set limit, the process is examined to find the source of the problem. Lower control limits for a P-chart are often ignored since this condition is usually desirable. See the "Displaying P-Charts" later in this topic.

The upper and lower limits placed on the various control charts can be devised by whatever makes sense to the process being measured. However, there are some common ways to choose these limits. Recall that a collection of quantitative observations has a distribution, and if those numbers have a normal (Gaussian, bell-shaped) type distribution, a measure for spread is the standard deviation. Thus, when data are normally distributed, (or if the sample sizes are large enough so the X-bars are approximately normal) you can calculate confidence intervals based on the standard error of the mean. An interval that is one standard error above and below the mean includes the true mean about 68.3% of the time. An interval of two standard errors includes the true mean about 95.4% of the time, and an interval of three standard deviations includes the true mean about 99.7% of the time. To devise lower and upper limits, most programs (including WINKS) use the "three-sigma" (three standard errors) criteria. That is, the limits are chosen so that there is about a 99.7% chance that the true mean is within the limits. If a mean is observed outside these limits, it is a "rare" event, and signals an out of control situation. Limits for R-Charts, S-Charts and P-Charts are also devised with the 3-sigma criteria. In the WINKS program, if you do not want to use the 3-sigma limits, you are given the option to enter your own upper and lower limits for the charts.

Note: Control limits for an X-Bar and P-Chart can change from sample to sample depending on sample size.

Note: The term "in control" means in "statistical control". That is, a process should not be called in control or out of control based on the statistical calculations unless the limits make sense for the item being measured. Statistical control relates to points occurring between some mathematically calculated limits, and may or may not have a relationship to reality.

X-Bar and R-Charts
(Analyze/Quality Control/X-Bar and R-Charts)

The data for X-Bar or R-Charts should be stored in a database using the following criteria. The data that will be used to calculate the means to be plotted come from a sample of observations, with each sample containing a number of replicates. For example, you might take samples of jars filled with jelly 25 times during the day. Each time you take a sample, it consists of 3 jars. Thus, you have

25 samples, each with a size of 3 (3 replicates). The data for this chart would be stored in a database using the following setup:

```
Observation   Sample   Value
1               1       15.9
2               1       16.1
3               1       16.0
:               :       :
4               2       16.2
5               2       15.9
:               :       :
6               3       15.6
etc.
```

Each jar should contain 16 ounces of jelly. You do not want the jars too empty or too full. Thus, you may want to see that the average amount of jelly does not go under or over certain limits. Also, you do not want the range to be too wide -- which may mean that the "average" jar contains 16 ounces, but the amount in different jars may vary widely. The database needed for this analysis would contain 2 fields, SAMPLE and VALUE. As an example, consider another problem: A manufacturer of automobile piston rings measures the diameter (in millimeters) of the piston rings to track the accuracy of the process. From 3 to 5 pistons were taken from 25 samples. The database to perform this analysis must contain at least two fields, SAMPLE and OBSERVED. The data are from Montgomery (1991, p 237). The data are in the file PISTONS.DBF. A partial listing of the data is:

SAMPLE	OBSERVED
1	74.030
1	74.002
1	73.992
1	74.008
2	73.995
2	73.995
etc.	
25	74.013

To analyze this data, follow these steps:

Step 1: Open the database named PISTON.

Step 2: From the Analyze menu, select "Quality Control Charts" then "X-Bar and R-Chart)".

Step 3: Choose the variable SAMPLE as the group variable and OBSERVED as the observation variable. Once you specify what fields to use, WINKS will display a preliminary screen showing the grand mean (X double-bar) and grand range (R-bar) and containing an options menu. See the figure below.

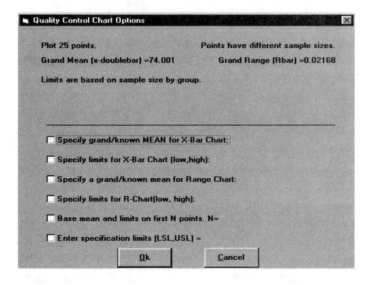

From this options menu you can select to display the control chart, specify a custom mean and limits, or view/print a listing of the computations for the control limits. When you continue, the results of the analysis are displayed in the viewer. Click on Graph to display the X-Bar control chart.

Click the Option button to select options such as axes labels, whether or not to display control limits and the R-Chart.

It turns out that the X-Bar chart reveals that the process is in
control since no points cross the upper or lower limits. Notice the
upper and lower limits differ across the range of the samples.
When the sample size is smaller, the limits are wider. This makes
sense because with more data, you have a better estimate of the
mean. When you choose the option to view the results of the
calculations, you can examine the values for the mean and range at
each sample and the upper and lower limits for the mean and
range. See the figure below.

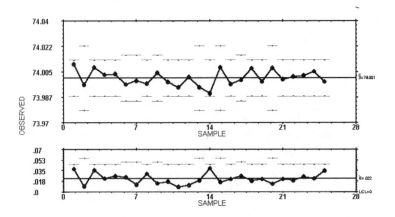

The tables output by the quality control module include the
numbers used to create the control charts. When an "*" asterisk
appears by a record in a table listing of this data, it means that the
point falls outside the control limits. This can help you quickly
identify points that are out of control. In the X-bar procedure, an
option has been added to allow you to specify lower and upper
specification limits. If you select this option Cp and Cpk
calculations are performed and reported in the output table.
(Similar to the calculations in the Detailed Cp,Cpk option.)

A technique of analyzing variability relative to product
requirements has been added to version 4.5 called process
capability analysis. In this analysis, Process Capability Ratios

(PCR) are calculated. For this analysis, the following terms are equivalent:

PCR = Cp (Process Capability Ratio)

PCRk = Cpk (one-sided PCR for specification limited nearest the process average)

(Cp and Cpk are commonly used Japanese terms for process capability calculations.)

For example, in the PISTONS example, once the preliminary information is displayed, enter your lower and upper limits at the line labeled "Enter Specification Limits." For this example, enter 73.95 and 74.05. The following information will be displayed at the bottom of the output (edited):

```
Process-Capability Calculations:
Specifications: LSL = 73.950 USL = 74.050
Estimated Process Sigma = 0.00979 (Using R-bar/D2 method)
Process capability ratio (PCR) Cp =1.7034
Percentage of tolerance band that the process (6 sigma)
uses up (P)= 58.71%
PCR(Low) =1.7434 PCR(High) =1.6633
PCR(k) (one-sided), Cp(k) =1.6633
```

X-Bar and S-Charts
(Analyze/Quality Control/X-Bar and R-Charts)

This is the same as the previous example, except the S-Chart, appears at the bottom of the screen instead of the R-Chart.

EWMA Calculations
(Analyze/Quality Control/EWMA)

This calculates the Exponentially Weighted Moving-Average Lambda, which is often used when it is desirable to detect out-of-control situations very quickly. EWMA calculations incorporate information from all previous subgroups, weighting the information from the closest subgroup with a higher weight.

P-Charts
(Analyze/Quality Control/P-Charts Equal or Unequal Subgroups)

P-Charts plot a proportion of items observed from within a sample. For example, you might take a sample of 25 items from a manufacturing process each hour. Then you count the number of defects in that sample. You are interested in plotting the proportion of defects across time to observe if an unusually high number of defects begin to occur. The format for the database is:

```
Sample       Size  Splits  Holes
1             30      1      5
2             30      4      1
3             30      2      2
etc.
```

For example, a large lumber yard makes a great deal of its profits from the sales of plywood. Recently, customers have been returning plywood sheets and asking for refunds. The customers complain that the sheets contain too many defects such as splits or holes. The owner of the lumber yard decided to carry out an investigation by looking at 150 plywood sheets that are produced each day for 20 days and recording the number of sheets that contained splits or holes. The data is in the database called PLYWOOD.DBF. To perform this example, follow these steps:

Step 1: Open the database named PLYWOOD.DBF.

Step 2: From the Analyze menu, select "Quality Control Charts" then "Proportions Chart (P-Chart)".

Step 3: Choose SAMPSIZE as the sample size variable and SPLITS and HOLES as the defect count fields.

Step 4: Click on Graph to display the p-chart. The p-Chart is shown in the figure below.

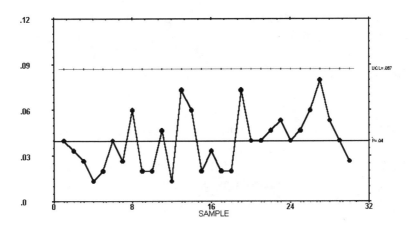

The SPLITS field contains the number of splits observed and the HOLES field contains the number of holes. Thus, for sample 1 the proportion of total defects found is 6/30. Using this same database, you can also create a P-Chart that only considers defects of type SPLITS. In this case, you would choose only the Sample Size and SPLITS fields for analysis.

The minimum number of fields in the WINKS database needed for this chart is two, a Sample Size field and a Count field. If there are more than one Count (defects) fields, the program will add up the defect fields to calculate the proportion defects for that sample. You do not need a SAMPLE field, but you may want one if the field will contain information about the sample, such as the hour taken.

In the plywood example, when the P-Chart for the total defects is plotted, the chart is in control with an average defect rate of about 4%. If this is satisfactory to the lumber yard then production may continue. If not, the lumber yard owner must decide to take steps to reduce the defect rate. Note that even though the chart that contains both types of defects is in control, if you plot separate charts for splits and holes, it is discovered that the splits chart is in control but the one for holes is not. This might indicate to management that attention should be given to reducing this particular type of defect (holes). Also, the manufacturer may want to set an upper limit for defects (say 0.06). If this were done, then the original p-Chart would show several instances of being out of control.

Notes on Analyzing Control Limits

The following items are adapted from The General Electric Handbook (1956) that contains suggestions for deciding when a process is out of control:

1. If one point is beyond the 3-sigma control limits.
2. If two out of three consecutive points are beyond 2-sigma limits.
3. If four of five consecutive points are beyond a 1-sigma limit.
4. If 8 consecutive points are on one side of the center line.

Some additional tips for analyzing control limits are:

If a single point in a plot is out of control - Check the original data to verify the correctness of this point. The problem may have been caused by an incorrectly recorded number.

If there is a trend in the control points - If control points in a plot are trending, up or down, it might indicate a problem such as a machine slowly getting out of calibration. Noticing this trend might help you locate the problem and make a correction before the process gets out of control.

If there is a pattern of several points in control, then out, then in, and so on - This might signal a step change in a process, such as a machine tool with a loose component that sometimes slips out of normal operation, then returns to normal operation.

Control Charts for Individual Measurements

In a situation where the sample size used in process control is 1, a
Control Chart for Individual Measurements is appropriate.

The result is similar to the X-Bar chart described in the previous
X-Bar and R-Chart example. A database on disk named PAINT
contains information about the viscosity for aircraft primer paint
(Montgomery, 1991, p. 242). The control Chart for PAINT is
shown in the figure below.

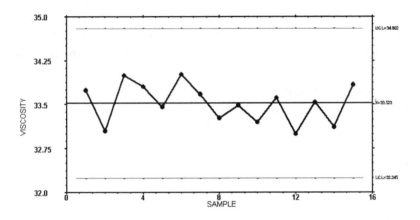

Pareto Charts Analysis
(Analyze/Pareto Charts)

The Pareto chart is a specialized bar-chart of categorical (or
descriptive) data in rank order. The usefulness of the Pareto chart
depends on the selection of the data for analysis. The data selected
should represent a characteristic of the organization (or system)
that is of interest to management. The items displayed in the chart
are arranged in decreasing order by frequency of occurrence,
allowing the user to see which items occur most frequently. Pareto
Charts can make it easy to interpret occurrences and their
importance as related to a process. Pareto Charts are a commonly

used Quality Management tool in both manufacturing and service industries. WINKS allows you to read in Pareto Chart data in two ways:

1. Read data, calculate frequencies, display plot - In this case, WINKS reads raw counts from a database similar to the frequency procedure. It then arranges the frequency categories in descending order to create the Pareto Chart.

2. Read frequencies, display plot - In this case, WINKS reads in frequencies that have already been tabulated. The frequencies are arranged in descending order to create a Pareto Chart. WINKS can also create charts using a "By" variable. This variable allows you to examine charts by some grouping factor such as day of week or operator, etc. When the chart is displayed, you can move from one chart to the next by choosing a next option from the chart menu. In the PARETO.DBF database there are two fields, named FAILURE and OPERATOR as shown here:

```
FAILURE    OPERATOR
Drift         1
Drift         1
Reagents      1
Reagents      1
  :    :
etc.
  :    :
Tubing        3
```

To create a series of three Pareto Charts, follow these steps:

Step 1: Open the database named PARETO.DBF.

Step 2: From the Analyze menu, choose "Pareto Charts"

Step 3: Select the FAILURE field as the Data field and the OPERATOR field as the Group (By) field.

Step 4: The Pareto chart will be displayed as in the figure below.

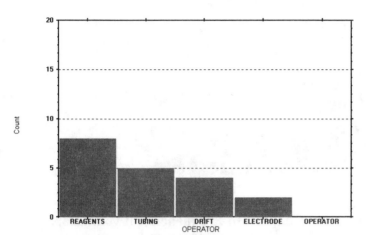

Notice on this plot the height of the bars are in descending order. Each bar is identified by a key that was the item counted from the database. The plot displayed is the one for OPERATOR=1. However, you can quickly display the other plots in one of two ways. First, notice the "next" button on the plot menu at the top of the screen. When you choose next, the plot for OPERATOR = 2 will be displayed. Or, you can exit the plot, and the "ABC" menu of plots will again be displayed. If there are a few plots, the "next" technique will be the fastest way to move from plot to plot. If there are many plots, it may be quicker to return to the plot menu, then select another plot to display.

You can choose Options on the menu to change titles, and select options for displaying the plot. From the options menu, you can choose the following three options Plot Curve (Y/N): This curve indicates the cumulative count or percent of failures from left to right. Display Counts (Y/N): This option causes counts to appear for each bar and for the Pareto curve, if it is present. Y-Axis as Percent (Y/N): This displays the y-axis as percents rather than counts. See the figure below.

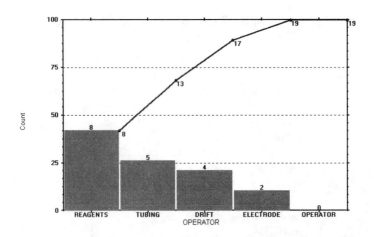

To Display Pareto Chart from Frequencies: If your database contains frequency counts, display a Pareto Chart by choosing the "Read Frequencies, display plot" option. For example, consider the database PARETOF.DBF:

LABEL	COUNT
STITCH	24
GRIND	32
LAYERING	12
WELD	8
LOOPS	41

Step 1: Open the database named PARETOF.

Step 2: From the Analyze menu, select "Pareto Chart from Frequencies"

Step 3: Select the COUNT field as the Data field, LABEL as the Label field.

Step 4: A menu appears containing options to plot the chart or quit. Choose plot options in the same way as described in the previous example.

References

Belsley, David A., Kuh, Edwin, Welsch, Roy E., *Regression Diagnostics: Identifying Influential Data & Sources of Collinearity*, Wiley, 1980.

Box, Jenkins, and Reinsel, *Time Series Analysis - Forecasting and Control*, Prentice-Hall, 1994.

Dallal, Gerard E. and Wilkinson, Leland, "An Analytic Approximation to the Distribution   of   Lilliefor's   Test Statistic for Normality," The American Statistician, Nov. 1986, Vol 40., No. 4, p 294-296.

Deming, W.E., Out of Crisis, Cambridge, MA: Massachusetts Institute of Technology, Center for Advanced Engineering Study, 1986

Dixon, W.J. and Massey, F.J., *Introduction to Statistical Analysis*, McGraw-Hill Book Company, New York, 1969.

Elliott, A.C. and Woodward, W.A.,"Analysis of an Unbalanced Two-Way ANOVA on the Microcomputer", *Communications in Statistics*, Volume B15, Number 1, 1986.

The General Electric Handbook, 1956.

Gunst, R.F., and Mason, R.L., *Regression Analysis and its Applications*, Marcel Dekker, New York, 1980.

Granger, C.W.J. and Newbold, P., *Forecasting Economic Time Series*, Academic Press, 1977.

Hoaglin, D.C., Mosteller, F., Tukey, J.W., *Understanding Robust and Exploratory Data Analysis*, John Wiley & Sons, Inc. New York, 1983. (Box and Whiskers Plots, Stem and Leaf Displays)

Kennedy, W. J., Jr., and Gentle, J.E., *Statistical Computing*, Marcel Dekker, Inc, New York, 1980.

Larsen, R.J. and Marx, M.L., *Statistics*, Prentice-Hall, 1990.

Lehmann, E.L. *Nonparametrics: Statistical Methods Based on Ranks*, Holden-Day, Inc, Oakland, Ca, 1975.

Lilliefors, Hubert, W., "On the Kolmogorov-Smirnov test for normality with mean and variance unknown", JASA, June, 1967, Vol 62. p. 399-402.

Longley, J.W. "An appraisal of least squares programs for the electronic computer from the point of view of the user." JASA, 1967, 62, 819-831.

Montgomery, D.C., *Introduction to Statistical Quality Control*, John Wiley and Sons, 1991.

Neter, J., Wasserman, W., and Kutner, M. H., *Applied Linear Statistical Models*, Richard D. Irwin, Inc., 1990, Third Edition.

Prentice, R.L. "Exponential survivals with censoring and explanatory variables.", Biometrika 60, 1973, 279-288.

Ryan, Thomas P. *Statistical Methods for Quality Improvement*, John Wiley & Sons, New York, 1989.

Tukey, J.W., *Exploratory Data Analysis*, Addison-Wesley, 1977.

Tsay, R.S., and Tiao, G.C., "Consistent Estimates of Autoregressive Parameters and Extended Sample Autocorrelation Functions for   Stationary and Nonstationary ARMA Models," JASA 79, 84-96, 1981.

Winer, B.J., *Statistical Principles in Experimental Design*, Second Edition, McGraw-Hill Book Company, 1971.

Woodward, W.A., Elliott, A.C., Gray, H.L and Matlock, D.C., *Directory of Statistical Microcomputer Software*, Marcel Dekker, New York, 1988.

Woodward, W.A., and Gray, H.L., "On the relationship between the S-array and the Box-Jenkins Method of ARMA Model Identification," JASA 76, 579-587, 1981.

Zar, J.H., *Biostatistical Analysis*, Prentice Hall, Inc, 1974 and 1984 editions.

INDEX

123 File Import 36
2x2 tables 159
All Subsets Regressions...182
Alpha Level 72, 100
Alternative Hypothesis 68
Analysis of Variance........See
    ANOVA
ANOVA.................66, 77, 97
Append.............................37
Area Chart........... See Graphs
ARMA ....See Time Series
    Analysis
ASCII............................44, 46
Autocorrelations 195
Autoregressive 192
Average............................52
Bar Chart........24, 60, 65, See
    Graphs
Best Subset Regression....182
Binary 185
Bland-Altman Plots 189
Box and Jenkins...............194
Box and Whiskers Plot 54
By Group 84
By Group Plots ... See Graphs
Categorical.........................75
Censor..............................143
Central Limit Theorem 148
Character............................31
Characteristic Equation....194
Chi Square 130
Chi-Square 77, 135
Cochran's Q 77, 115

Coefficient ....See
Cohen's Kappa...See KAPPA
Collinearities....175, 177, 183
Comma Delimited..............46
Comparing Means..............13
Confidence Interval 80, 81,
    147
Confounding 161
Contingency Coefficient..136
Contingency Coeffieicent..78
Control Chart for Individual
    Measurements..............213
Control Charts..................213
Control Limits..................212
Correlation 60, 61, 64, 78,
    117, 124, See Graphs
Correlation Coefficient 119
Correlation Matrix 126
Counts 84, 150
Cp....................209, See Cpk
Cpk...........................82, 209
Cramer's V 136
Create a Database 29
Critical Region...................71
Crosstabulation 133
Crosstbulation 76
Cubic...............................175
Curvilinear 174
Customized Database..30, 32,
    See Database
Data Report.......................47
Database...........................134
dBase...... See DBF, See DBF

DBF ....See dBase
Decimals............................ 32
Decision Tree 75
Degrees of Freedom 71
Delete 41
Delete a Database.............. 35
Delete Row........................ 15
Demming.......................... 202
Demonstrations 147
Descriptive Statistics......... 76
Dichotomous ... 115, 129, 140
Differencing 200
Discriminant Analysis....... 78
Dispersion 52, 58
Display Structure............... 37
Dunnett's Test.................. 108
Edit Menu.......................... 36
Equality of Variance 97
EWMA 209
Excel........ 26, 36, 44, 45, See
    Importing
Exponentially Weighted
    Moving-Average 209
Export............................... 35
F Distribution 70
Fieldname.......................... 31
File Menu 35
First Impression......... 22, See
    Graphs
Fisher's Exact Test 136
Fixed............................... 165
Flip a Coin....................... 147
Forecast 198, 201
Format Definitions 158
Formula 38, 42
Frequencies 130
Friedman 77, 113

Genetic 131
Goodness-of-Fit 77, 131
Graphical Correlation Matrix
    128
Graphical Interpretation 17
Graphs 88
Grouping Variable............. 34
Help 9, 35
Heteroscedasticity 124
Histogram.............. 19, 57, 88
Homogeneity 139
Hypothesis........................ 68
Import........................ 35, 44
Independent 75
Independent Group t-test.. 14,
    See t-test, See t-Test
Indicator Variables.......... 180
Individual Measurements 203
Installation........................ 10
Interaction 166, 172
Interaction Plots 168
Intercept 119
InterRater Reliability....... See
    KAPPA
Journal 36
Kaplan-Meier .... See Survival
    Analysis
Kappa 162
Kendall Partial Rank 78
Keyboard......................... 133
Kolmogorov-Smirnov 80
Kruskal-Wallis 77, 110
Lambda............................ 209
License ii
Life Table .... See Survival
    Analysis
Limitations 35

Line Chart .... See Graphs
Linear 175
Linear Regression.See
 Regression
Logical 31
Logistic 185, 188
Logit 186
Longley 122, 184
Lotus 1-2-3 44
Mallow's Cp 182
Mann-Whitney 77, 108
Mantel-Haenszel 144, 159
Matrix of Scatterplots 22
McNemar 67, 77, 115, 140
Mean 52
Mean Rank 109
Median 54, 80
Memo fields 35
Missing Values 37, 40, 151
Model Identification .... See
 Time Series Analysis
Modeling 195
Moving Average 192
Multicollinearity 124
Multiple Comparison .. 67, 73,
 100, 167
Multiple Regression... 78, See
 Regression, See
 Regression
New Database 35
Newman-Keuls 74, 99, 105
Non-Parametric 108, 112
Nonstationary 194, 199
Normal 56, 75, 81
Normal Probability Plot 82
Null Hypothesis 67
Numeric 31

Odds Ratio 136, 137
One-Way ANOVA .... See
 ANOVA
Open a Database 35
Outliers 180
Paired .... 102, See Repeated
 Measures
Pareto Chart 213
Parsimony 124, 192
Partial Autocorrelations ... 195
Paste 38
P-Chart 203, 210
PCR 209
Pearson.. See Correlation, See
 Correlation
Pearson Correlation .... See
 Correlation
Percentile 54, 81, 84
Phi Coefficient 136
Pie Chart .... See Graphs
Point Bi-Serial .... See
 Correlation
Polynomial Regression .... See
 regression
Pre-Defined Structure 30
Predict 120, 123
Print 36, 37
Probability 87
Process Capability Ratios 208
Proportions 141, 142
p-Value 73, 86
Qualitative 59
Quality Control 202
Quantitative 52
R^2 ..... 119, 125, See R-Square
Random 165
Random Sample 39, 41

R-Chart.................... 203, 205
Recode.......................... 39, 43
Regression... 60, 63, 116, 174
Regression Equation 178
Regression Through Origin
  120
Rejection 71
Related 75
Relative Risk 136, 137
Remove Formulas 38
Repeated Measures .. 77, 113,
  See ANOVA, See
  ANOVA, See Paired
Residuals 124
R-Square.......................... 124
Runs Test.......................... 77
S.E.M 81
Scatterplot ... 62, 88, 117, See
  Graphs
S-Chart 203, 204, 209
Scheffe 172,
Sensitivity............... 136, 138
Sign test.......................... 112
Significance Level............. 69
Simple Linear Regression See
  regression
Simple Logistic Regression
  .... See Logistic
Simulations...................... 147
Single Sample t-test 106, 107
Sort 39, 41
Spearman..... See Correlation,
  See Correlation
Spearman's Correlation.... 62,
  See Correlation
Specificity 136, 138
Spreadsheet 14, 36

SQC 202
Standard Deviation 54, 80, 81
Standard Error of the Mean 80
Stationary 199
Stem and Leaf 76, 85
Stepwise ....See Regression
Student's t-test.................. 11
Student's t distribution....... 70
Summarized Data.............. 12
Survival Analysis .....See Life
  Table
Tabulation 150
t-Distribution..................... 70
Text Replace 39
Three-Way ANOVA .... See
  ANOVA
Time Series 76
Time Series Analysis 192
Transformations 180
t-test................................. 94
t-Test 106
Tukey 172
Tukey 5-Number Summary
  82
Tukey Multiple
  Comparison............. 111
Tutorials 11
Two-Way ANOVA .... See
  ANOVA
Two-Way Repeated Measure
  .... See ANOVA
Two-Way Unbalanced 168
Unbalanced Design .... See
  ANOVA
Updates.......................i, 224
Utilities............................ 35
Variance 80

Wilcoxon.....................77, 112
W-statistic....See Time Series
  Analysis

X-Bar chart 203
X-Bar Chart 205
Yates' Chi-Square 136

*For updates, errata and other
information concerning
this manual or program go to*

http://www.texasoft.com/updates